Immunohistochemistry and Immunocytochemistry

Immunohistochemistry and Immunocytochemistry

Essential Methods

EDITED BY

SIMON RENSHAW

Second Edition

WILEY Blackwell

This edition first published 2017 © 2017 by John Wiley & Sons, Ltd.

First edition published 2005 by Scion Publishing Ltd.

Registered office: John Wiley & Sons, Ltd, The Atrium, Southern Gate, Chichester, West Sussex, PO19 8SQ, UK

Editorial offices: 9600 Garsington Road, Oxford, OX4 2DQ, UK
The Atrium, Southern Gate, Chichester, West Sussex, PO19 8SQ, UK
111 River Street, Hoboken, NJ 07030-5774, USA

For details of our global editorial offices, for customer services and for information about how to apply for permission to reuse the copyright material in this book please see our website at www.wiley.com/wiley-blackwell.

Library of Congress Cataloging-in-Publication Data applied for.

9781118717776

A catalogue record for this book is available from the British Library.

Wiley also publishes its books in a variety of electronic formats. Some content that appears in print may not be available in electronic books.

Cover image: © Abcam plc

Set in 10/13pt, PhotinaMTStd by SPi Global, Chennai, India.

1 2017

FSC www.fsc.org MIX Paper from responsible sources FSC® C013604

Table of Contents

List of Contributors vii

Preface ix

Acknowledgements xi

Chapter 1: Antibodies for Immunochemistry 1

Mark Cooper and Sheriden Lummas

Introduction 1
Immunogens for Antibody Production 5
Antibody Production 12
Antibody Purification 16
Fragment Antibody Preparations 20
Antibody Labelling 21
Antibody Stability and Storage 23
References 24

Chapter 2: The Selection of Reporter Labels 25

Judith Langenick

Introduction 25
Enzymatic Labels 26
Fluorescence Detection 29
References 32

Chapter 3: Immunohistochemistry and Immunocytochemistry 35

Simon Renshaw

Specimen Formats for Immunochemistry 36
Fixation 37
Processing Tissue Blocks to Paraffin Wax 46
Microtomy 47
Tissue Microarrays 47
Specimen Storage 48
Decalcification 49
Antigen Retrieval 50
Controls 56

Immunochemical Staining Techniques (Optimizing a New Antibody) 57
Counterstains 71
Mounting 74
Troubleshooting 76
Examples of Immunostaining Photomicrographs 76
Acknowledgements 101
References 101

Chapter 4: Multiple Immunochemical Staining Techniques **103**
Sofia Koch

Introduction 103
Methods and Approaches 115
References 122

Chapter 5: Quality Assurance in Immunochemistry **123**
Peter Jackson and Michael Gandy

Introduction 123
Methods and Approaches 125
Automated Immunochemical Staining 147
Troubleshooting 149
References 154

Chapter 6: Automated Immunochemistry **157**
Emanuel Schenck and Simon Renshaw

Introduction 157
Methods and Approaches 160
Other Forms of Automation 164
References 168

Chapter 7: Confocal Microscopy **169**
Ann Wheeler

Introduction 169
When Should Confocal be Used? 173
Applications: For Example Co-localization, Quantification, 3D Visualization and Kinetics 173
How To Set Up a Confocal Experiment? 174
References 198
Further Readings 198

Chapter 8: Ultrastructural Immunochemistry **199**
Jeremy Skepper and Janet Powell

Introduction 199
Methods and Approaches 207
References 222

Index **227**

List of Contributors

Mark Cooper
Abcam plc, Cambridge, UK

Michael Gandy
The Doctors Laboratory Ltd, London, UK

Peter Jackson
Department of Histopathology, Leeds General Infirmary, Leeds, UK (retired)

Sofia Koch
Abcam plc, Cambridge, UK

Judith Langenick
AbD Serotec, Oxford, UK

Sheriden Lummas
Abcam plc, Cambridge, UK

Janet Powell
Cambridge Advanced Imaging Centre, Department of Anatomy, University of Cambridge, Cambridge, UK

Simon Renshaw
Abcam plc, Cambridge, UK

Emanuel Schenck
Medimmune LLC, Gaithersburg, MD, USA

Jeremy Skepper
Cambridge Advanced Imaging Centre, Department of Anatomy, University of Cambridge, Cambridge, UK

Ann Wheeler
Institute of Genetics and Molecular Medicine Advanced Imaging Resource, University of Edinburgh, Edinburgh, UK

Preface

IMMUNOCHEMISTRY IS AN INVALUABLE TOOL for the visualization of cellular antigens in diagnostic and biological research environments. The need to obtain accurate, reliable and reproducible results is of paramount importance.

It is with this fundamental aim in mind that we have compiled *Immunohistochemistry and Immunocytochemistry: Essential Methods*. We have achieved this by examining each aspect of immunochemistry in turn, with each chapter including detailed information regarding the subject matter in question. Each chapter is written by an expert in their field and includes protocols that are typically used in their own research. In addition, benefits and limitations of each approach are discussed within the chapters.

This book offers a wealth of knowledge to the novice immunochemist, who, from the outset, wishes to fully understand the theory and practice of immunochemical staining techniques and obtain reliable and reproducible data time and time again. For the experienced immunochemist, this book is a comprehensive reference guide to the theory and practice of immunochemical staining techniques, allowing further optimization of existing immunochemical staining protocols.

Simon Renshaw
January 2017

Acknowledgements

THANK YOU TO ALL of my friends, family and colleagues for your continued support throughout this project.

A special thank you goes to the contributing authors, without whom this book would have taken considerably longer to write!

Thank you to Elsevier Ltd, Abcam plc and Leica Biosystems for kindly agreeing to reproduction of copyrighted materials.

Finally, a very special dedication goes to Chris van der Loos, who had very kindly agreed to be the author of the 'Multiple Immunochemical Staining Techniques' chapter, but sadly passed away before beginning the work. He was incredibly gifted in his field and delivered a most informative and entertaining lecture. He will be missed by many.

1

CHAPTER ONE

Antibodies for Immunochemistry

Mark Cooper and Sheriden Lummas

Abcam plc, Cambridge, UK

INTRODUCTION

Unlike innate immunity, the adaptive immune response recognizes, reacts to and remembers foreign substances invading an organism. Antibodies play a central role in the function of adaptive immunity. Their roles are to detect, specifically bind and facilitate the removal of foreign substances from the body. Memory B cells create an immunological memory that allows the immune system to respond quicker upon subsequent exposure to the same foreign substance.

A substance not recognized by the immune system as being native to the host and therefore stimulates an immune response is known as an antigen (antibody generator). Binding of an antigen to an antibody is specific. Biochemical research utilizes the ability of antibodies to distinguish between antigens and to detect biological molecules (commonly proteins) in cells and tissues using immunochemical staining techniques. Immunochemistry is the focus of this text, and its practice is discussed in detail throughout later chapters (see p 35).

Immunohistochemistry and Immunocytochemistry: Essential Methods, Second Edition. Edited by Simon Renshaw.

Typical Antibody Structure

Antibodies are immunoglobulin (Ig) proteins produced by B cells in the presence of an antigen. Immunoglobulins exist as five main classes or isotypes: IgA, IgD, IgE, IgG and IgM. Each isotype performs a different function in the immune system. IgG has a long half-life in serum (Table 1.1), which means its clearance from the circulatory system is slow. The abundance and retention of IgG in circulation compared to the other classes make it the most common antibody isotype reagent used in biochemical research.

The basic antibody unit is shared across all five isotypes. Two identical heavy (H) and light (L) polypeptide chains connected by a disulfide bond form the commonly illustrated Y-shaped antibody structure (Fig. 1.1). The arms of the Y structure form the Fab (fragment antigen-binding) region while the base is the Fc (fragment crystallizable) region.

Both H and L chains consist of variable (V) and constant (C) domains, named according to the conservation of their amino acid sequence. One variable domain is present for each H chain and L chain and is situated at the amino terminus. V_L and V_H domains are paired together to create the antigen-binding site (paratope). Specificity of antigen

TABLE 1.1 A Comparison of Immunoglobulin Classes
IgG has the longest half-life of all the antibody classes and is produced during the secondary immune response. IgG, IgD and IgE are monomeric structures consisting of a single antibody unit. IgA can occur as a monomer or dimer (two units). IgM exists as a pentameric molecule, with five basic immunoglobulin units joined by an additional polypeptide chain (J chain), making it the largest antibody class with a molecular weight of 970 kDa.

	Immunoglobulin								
	IgG1	**IgG2**	**IgG3**	**IgG4**	**IgM**	**IgA1**	**IgA2**	**IgD**	**IgE**
Molecular weight (kDa)	146	146	165	146	970	160	160	184	188
Serum level (mean adult mg ml^{-1})	9	3	1	0.5	1.5	3	0.5	0.03	5×10^{-5}
Half-life in serum (days)	21	20	7	21	10	6	6	3	2
Location	Bloodstream. Can pass through blood vessel walls readily and cross into the placenta.				Bloodstream	Body secretions: tears, sweat, saliva; breast milk		B-cell surface	Bound to mast cells
Function	Activates complement pathway				Produced during primary immune response and activates complement system	Form a defence on the surface of body cells. Immune protection to newborn.		Unknown	Stimulates allergy response

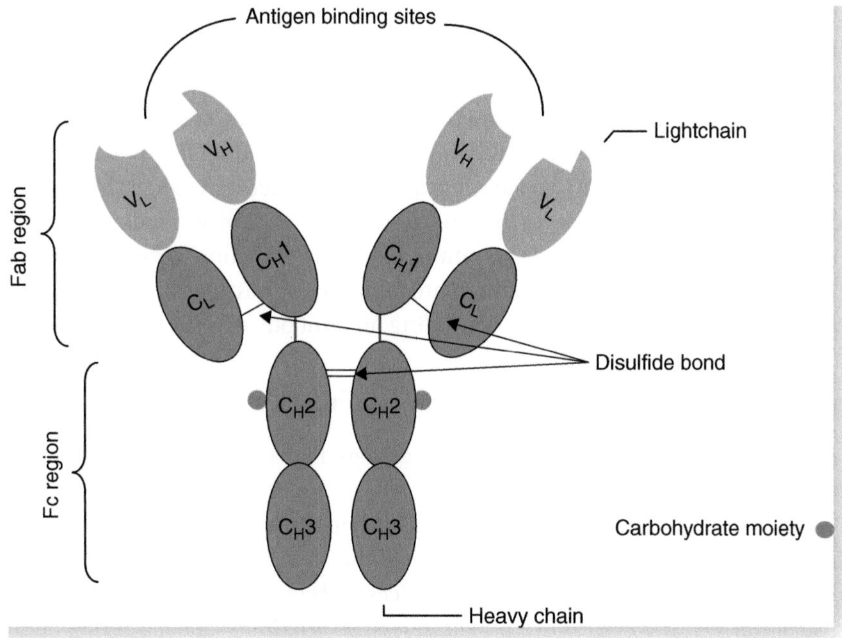

FIGURE 1.1 A Schematic Diagram of an Immunoglobulin Molecule
The basic antibody molecule is a Y-shaped structure consisting of two heavy and two light polypeptide chains joined by disulfide bonds. The chains are composed of variable (orange) and constant (blue) immunoglobulin domains. The antibody–antigen binding site (paratope) is located at the tip of the Y arms. The Fab and Fc regions of the molecule are tethered together by a hinge region, which provides the antibody molecule with flexibility. Constant domains share the same amino acid sequence for a given antibody isotype. Different immunoglobulin isotypes arise from subtle sequence variations in the constant domains.

binding is determined by the variation in amino acid sequence in this region. This enables antibodies to recognize a diverse range of antigens, even though only a single amino acid difference between antigens exists. The remainder of the H and L chains are composed of constant domains: one domain in the L chain and three domains in the H chains denoted as C_H1, C_H2 and C_H3. For a given isotype, the entire C_H amino acid sequence is conserved. Subtle differences in the sequence occur and give rise to isotype subclasses (as provided in Table 1.1). IgG sub-classes are present across species, with IgG1, IgG2, IgG2a and IgG3 subclasses existing within mice. The subclasses have different binding affinities for the purification of protein resins, which are discussed in the 'Antibody Labelling' Section. It is therefore important for the sub-class to be accurately determined in order to efficiently purify antibodies from sera. Furthermore, primary antibody subclass determination is one of the critical parameters (among others) that enables the end user to select an appropriate secondary antibody.

Differences in C_L sequence equate to the type of L chain, out of which two types are found in antibodies. L chains exist as lambda (λ) or kappa (k) and are identically present in one form or the other in a single antibody. At the centre of the Y-shaped structure is the hinge region that acts as a tether, linking the Fab and Fc regions of the molecule. The two Y arms of the Fab region are able to move independently, providing the molecule with flexibility when the antibody binds two identical antigens, particularly when the antigens are distances apart [1]. This is a property that contributes to the use of antibodies in immunoassays. Being glycoproteins, antibodies contain a sugar side chain (carbohydrate moiety). This is bound to the C_H2 region and contributes to antibody destination within tissues and the type of immune response initiated depending on antibody class [1].

Antibody Structure Is Optimized for Its Function

In order to understand how antibodies are engineered for their function, an overview of protein structure organization has been presented.

Proteins are composed of polypeptide chains consisting of basic units called amino acids. The consecutive sequence of amino acids is the primary structure. Hydrogen bonds within the polypeptide chain generate alpha helices and beta sheets to create the secondary protein structure. As the chains are pulled into close proximity of each other, additional bonds and interactions form between amino acid side chains, and hydrophobic bonds and van der Waals interactions form between non-polar amino acids. Further reinforcement of the conformational structure is achieved by the formation of disulfide bonds between cysteine sulfhydryl groups. The result of these bonds is the generation of polypeptide subunits, that is, the tertiary protein structure. The arrangement into multi-subunit structures creates the final quaternary protein structure [2]. Two main types of quaternary protein structure exist: globular and fibrous.

Immunoglobulins are a superfamily of globular proteins with roles associated with the immune system. Examples include cell surface receptors (Fc) and antibodies. Members of this superfamily exhibit a common structural motif, that is, the immunoglobulin domain [2]. The immunoglobulin domain is approximately 110 amino acids in length. Two immunoglobulin domains are present in the light chain, whereas the heavy chain has four domains, numbered from the amino (N-) terminus to carboxyl (C-) terminus. Each domain is a sandwich-like structure formed from anti-parallel beta-pleated sheets of the polypeptide chain bound together by disulfide bonds. This structure is known as the immunoglobulin fold. Loops are created at the ends of the immunoglobulin folds where the beta-pleated sheets change the direction. These loops are 5–10 amino acids residues in length and reside within the variable regions, protruding from the surface. They are designated hypervariable (H_V) loops because the amino acid sequence variation within this region is considerable. H_V loops are also referred to as complementarity-determining regions (CDRs). The three H_V loops present in the variable region are denoted H_V1, H_V2 and H_V3, with H_V3 containing the greatest sequence variation. The remainder of the variable region is composed of framework regions FR1, FR2, FR3 and FR4, respectively. These regions lie between the H_V loops, have less sequence variation and provide structural integrity to the immunoglobulin molecule. Pairing of the H_V loops from the heavy and light chains in the antibody molecule creates a single antibody–antigen binding site at the tip of the Y arms. The amino acid sequence determines the tertiary structure of this site, described as the paratope. The surface of the paratope is complementary to a specific amino acid sequence on the antigen's surface (the epitope), and thus dictates antibody specificity. Pairing different combinations of V_L and V_H regions generates the diverse repertoire of antibodies [1].

Antibodies must be able to perform their biological function under broad and changing conditions. Intermolecular disulfide bonds occur between the heavy and light chains along the entire length of the antibody and contribute to the stability of the molecule. Out of academic interest, by the use of reducing agents such as dithiothreitol (DTT) and 2-mercaptoethanol, the disulfide bonds can be removed to denature the antibody into its heavy and light chain fragments, with molecular weights of 50 and 25 kDa, respectively. Sodium dodecyl sulphate-polyacrylamide gel electrophoresis (SDS-PAGE) using reducing conditions is routinely employed to resolve the antibody fragments.

IMMUNOGENS FOR ANTIBODY PRODUCTION

Protein Structure Considerations for Antibody Generation

The large globular structure of proteins means binding to an antibody is achieved through continuous or discontinuous epitopes. Antibodies raised using a full-length protein recognize a combination of amino acids brought together in the protein's three-dimensional conformation (tertiary structure). The amino acids are discontinuous, and this type of epitope is termed 'conformational'. Antibodies also recognize linear epitopes by the consecutive sequence of amino acids (primary structure) [3]. *In vivo* digestion of a foreign substance by macrophages yields small segments of its sequence. Commercially available antibodies raised against peptide immunogens mimic this biological process.

The likely success of an antibody recognizing a continuous versus conformational epitope generally cannot be predicted. Antibodies that recognize the native protein structure are more likely to perform well in immunochemistry, due to the protein being in a more *in vivo* state when compared to other immunoassay techniques such as western blotting, when performed under reducing conditions. However, fixation, tissue preparation and antigen retrieval methods can potentially have a detrimental effect on the tertiary protein structure by chemically modifying amino acids [4]. It is therefore recommended that researchers evaluate and optimize their antibody in a given application (see p 59). Antibodies purchased from commercial suppliers should be supported by characterization data. Often, the protocol employed to obtain the desired staining pattern will be provided or made available upon request.

Types of Immunogens for Antibody Production

The range of antigens (immunogens) able to facilitate an immune response is diverse and may originate from within an organism or from the external environment. Antigens employed to produce antibodies to specific protein targets in industry and academia include peptides, whole cells, nucleotides and recombinant proteins. However, peptides are the most commonly used antigen, which is represented by the vast number of custom peptide suppliers available commercially. Peptides offer two main advantages over other antigens. Firstly, they are simple and quick to synthesize. Secondly, cross-reactivity with related proteins can be minimized through considered selection. Furthermore, antibodies can be raised to specific post-translational modifications such as methylation, acetylation and phosphorylation, which are important for epigenetic advances. The main limitation of peptide immunogens is that antibodies generated are less likely to recognize the protein's native structure. If candidates are poorly selected, the likelihood of epitopes lying within a region that is not accessible (due to the tertiary protein structure) is greater. This may be seen in immunochemical assays as poor or no positive staining [4]. The following sections aim to alleviate this by presenting tools and guidelines for the determination of protein structure and the subsequent selection of an antigenic and immunogenic peptide immunogen. It should be noted that this information will only be of any real use to people who are designing peptide immunogens for antibody production. Most end users will simply purchase an antibody from a commercial supplier, where this will have already have been done. However, the information will help to give the end user a wider appreciation of how antibodies are designed and produced, and how immunogen design can either negatively or positively affect the performance of an antibody in any given immunoassay.

Epitope Prediction Tools

Traditional methods for predicting a protein's conformational structure and sites of antigenicity assigned a value to each amino acid based on their physiochemical properties to generate a scale of propensity [5]. Prediction of secondary protein structure based on amino acid sequence was one of the first propensity scales to have been developed by Chou and Fasman [3]. Hydrophilicity and antigenicity indices for epitope prediction followed shortly after. The bioinformatics information that is available today is vast and extends beyond the indices and scales sited in this text. A variety of new systems have been developed over the years alongside evaluations and adaptations of historical methods. Recently, epitope prediction resources in the public domain frequently combine the historical propensity scales with novel mathematical algorithms, an example being the BepiPred method to predict linear B-cell epitopes [3].

Developments in drug discovery have prompted bioinformatic advances for epitope prediction to generate antibodies with high affinity and specificity for therapeutic and diagnostic use. Historical systems provided a general protein model to predict antigenic locations. However, a method for determining antigenic sites specific to individual antibodies from epitope–paratope amino acid composition has recently been presented [6]. Furthermore, advances in antibody modelling are facilitating computerized antibody design [7].

Although these developments are a significant improvement for epitope prediction and offer time saving over peptide scanning experiments, further refinement is required before researchers become solely reliant on these models. In the following sections, the general principles are presented that provide a guide to the basic peptide immunogen design, utilizing public domain resources. Consideration of peptide candidates at this stage of antibody production can help generate a specific antibody for the immunochemistry end user. Custom antibody and peptide suppliers are widely available and may offer consultation for advanced immunogen design.

Considerations for Peptide Immunogen Design

Research the Protein

UniProt (http://uniprot.org) is a proteomic database providing protein sequence and function information. Comparing your protein sequence against the UniPort entry is a good practice. Information on isoforms, protein topology and post-translational modifications is provided alongside tissue expression and cellular localization, which will assist identifying a good peptide candidate.

For proteins highly conserved across family members or that contain isoforms, an alignment of the primary sequences is recommended to help avoid or minimize unwanted cross-reactivity. This will highlight regions of the sequence that are distinct from related proteins. Similarly, if cross-reactivity with a particular species is required for your experiments, performing a sequence alignment of the target and secondary species will identify conserved regions as candidates for further investigation. UniProt has alignment functionality. An additional alignment program available in the public domain includes Clustal Omega provided by EMBL-EBI (http://ebi.ac.uk).

It is highly advisable to record the expected size of the protein, tissue expression and subcellular localization. This information will be useful when comparing candidate immunogens against unrelated proteins for cross-reactivity. Cross-reactivity with proteins of similar size (and in particular for immunochemical assays, incorrect tissue expression and cellular localization) is concerning because doubt will be cast over the specificity of antibodies to the correct antigen. Care at the peptide selection stage will reduce the likeliness of cross-reactivity. However, there are proteins that are yet to be discovered and curated, so it is impossible to remove all the potential for cross-reaction. It is recommended in these instances to employ additional assays, for example Western Blot, where the proteins differ in molecular weight, to help verify specificity.

Identifying Candidate Regions

A requirement for a good peptide immunogen is for it to originate from an external exposed (hydrophilic) region of the protein, as this increases the chances of recognizing the tertiary protein structure. This information can be obtained from the tertiary protein structure. If tertiary structure is unavailable (e.g. for a novel protein), then hydrophilicity can be used as a measurement of potential surface exposure. Hydrophilic peptides ensure solubility, which is a prerequisite for synthesis and immunization. The Immune Epitope Database (IEDB) (http://iedb.org) is a public domain resource for B-cell epitope prediction providing hydrophilicity (Parker), secondary structure (Chou and Fasman) and BepiPred (Larsen) analyses for a given protein. Hydrophilic amino acids include serine (S), cysteine (C) and threonine (T) [2].

N- and C-terminal regions are likely to protrude through the surface of the protein, making them as good potential candidates. The C-terminal region is frequently not conserved between species. If species cross-reactivity is desired, performing a sequence alignment is essential before selection.

Protein topology is particularly informative for membrane proteins. A proteomic database will provide locations of transmembrane regions, which should not be selected for an immunogen, since this region is often inaccessible to an antibody. The same applies to cleaved regions, such as pro-peptides and signal peptides, since these will not be present on the mature form of the protein (unless, of course, you specifically require an antibody to recognize these). Instead, select candidates within extracellular domains or cytoplasmic loops. Selection of these regions mimics the presentation of the native protein. Therefore, the probability of generating antibodies suitable for a variety of assays is greater and increases the likelihood for the recognition of fixed protein, such as in immunochemical staining.

Specific Amino Acid Properties

Ideally, candidates should contain immunogenic residues such as proline (P) and tyrosine (Y), which provide a structural motif likely to be present in the native protein. The ring-like structure formed provides rigidity towards the N terminus. This provides greater exposure to the immune system when present inside the host organism compared to a coiled peptide. The position of these residues within the peptide sequence will dictate how well their property is translated.

Hydrogen bonds form between polar side chains of amino acids, such as glutamine (Q). This reduces solubility, and in the presence of too many hydrophobic residues it may cause the peptide to precipitate out of the solution. A good combination of hydrophilic and hydrophobic residues is preferred.

Glycine (G) is the most abundant amino acid. It is also the smallest and creates flexibility in the polypeptide chain. Conformational changes occur in the presence of multiple glycines across the peptide sequence. This should be avoided. Methionine (M) is usually the first amino acid in the protein sequence and can be cleaved off when the protein is processed into its mature form. Methionine can also undergo oxidation. For this reason, restrict the number of methionine residues within the peptide sequence if unavoidable. Cysteines (C) form disulfide bonds under oxidation and provide stability to the tertiary protein structure. Disulfide bonds between cysteine residues are common in structural proteins like keratin along with proteins that function in harsh environments, for example the digestive system. Consecutive serine (S), threonine (T), alanine (A) and valine (V) residues impede peptide synthesis. Avoid them where possible or consult a peptide supplier for a strategy to overcome the synthesis issues.

Review Potential Cross-Reactivity

Candidate peptides identified using the aforementioned guidelines must be checked for cross-reactivity with unrelated proteins, for reasons previously stated. Cross-reactivity with related proteins will already have been addressed by performing a sequence alignment. A Basic Local Alignment Search Tool (BLAST) algorithm, such as NCBI blastp (http://blast.ncbi.nlm.nih.gov/Blast.cgi), identifies and annotates locations of similarity between protein sequences. The precise nature of the epitope is unknown; therefore, gauging the potential for these regions to cross-react is purely subjective. As a guide, avoid five of more consecutively shared amino acids with unrelated proteins present in the species of interest, particularly proteins of similar molecular weight, tissue expression or localization. The BLAST results may identify cross-reactive regions within the peptide sequence, and reduction or elimination of these is possible by extending or shortening the sequence. Restricting the BLAST search to the organism of interest will aid interpretation of results. If species cross-reactivity is desired, then a BLAST against the species proteins is required to inform the final peptide selection.

Refining Selected Candidates

A length of 15–20 amino acids is common for peptide immunogens and allows for the generation of multiple epitopes across the peptide sequence. For small proteins, it may not be possible to implement the guidelines to the best ability and obtain a sequence of 15 residues. Consult with a peptide supplier in these situations and establish a strategy for increasing the overall length to facilitate an immune response without compromising antibody specificity.

Considerations for Post-Translational Modification Peptide Immunogens

Unlike unmodified peptide immunogens where the entire protein sequence is reviewed and analysed before a final peptide sequence is selected, the sequence used for a modified peptide is dictated by the site of the post-translational modification. The modified residue is usually positioned at the centre of the sequence to ensure that the epitope incorporates the modification. As a general guide, ensure that the modified residue is flanked by five amino acids on either side. Increasing the number of flanking residues beyond seven is not recommended, since the likelihood of the resulting antibodies recognizing an epitope lying in the flanking region is increased and, therefore, risks antibodies favouring the unmodified form of protein dominating the modified form. An unmodified version of the peptide sequence should be synthesized to remove from the sera (by affinity purification), any resulting antibodies that recognize the unmodified form, and to serve as a specificity control in immunochemical experiments.

Peptide Carrier Protein

Peptides are immunogenic, but the antibody response raised when used on their own is low, since they are typically around 15–20 amino acid residues in length (<6–10 kDa). Therefore, peptides are typically conjugated to a larger, immunogenic carrier molecule in order to generate a significant amount of antibody [8]. The process of peptide conjugation is discussed in more detail later in this chapter. Conjugation to KLH is commonly via a cysteine (C) residue in the peptide sequence. If a cysteine residue is absent, one must be added to either end of the sequence prior to peptide synthesis.

There are many types and variations of carrier proteins that are commercially available, such as ovalbumin (OVA) and catonized BSA (cBSA). However, the most popular and widely used carriers in antibody production are bovine serum albumin (BSA) and keyhole limpet hemocyanin (KLH).

BSA makes a suitable carrier protein due to a number of beneficial characteristics. It is stable and readily soluble in aqueous buffers. It is large enough in size (67 kDa) and sufficiently complex to be immunogenic. BSA has numerous carboxyl groups and primary amines, which are suitable for coupling peptides using a broad range of cross-linking reagents. A potential downside to the use of BSA as a carrier protein is that immunochemical experiments commonly use BSA as a reagent to block non-specific protein–protein interactions. Therefore, if antibodies that recognize the peptide–BSA complex are present in the final antibody preparation, false-positive/general background staining may occur [9].

KLH is an effective carrier protein for several reasons. It is large and complex ($2–3 \times 10^6$ Da) and has numerous epitopes and a wealth of lysine residues for coupling peptides. KLH is isolated from the mollusk *Megathura crenulata*, which is phylogenetically distant from mammals, ensuring a good host immune response [8]. This means that KLH shows a higher immunogenicity compared with BSA. A downside to the use of KLH is limited solubility in aqueous buffers. However, due to the popularity and extensive use as a carrier, commercial product variants (mcKLH) are available. These are purified, lyophilized and pre- activated forms, which are opalescent blue when in solution.

Apart from using carrier proteins to prepare peptide immunogens, an alternative is to synthesize peptides as multiple antigenic peptides (MAPs). MAPs are complexes created from multiple copies of the peptide immunogen coupled with a core structure via lysine residues. The advantage of MAPs is that as they do not require a carrier protein, antibodies are only generated against the peptide and not the highly immunogenic carrier protein. The use of MAPs is, however, limited to peptides that contain a good T-helper-cell epitope due to lack of affinity maturation and class switch to IgG. The purity of the MAP is also difficult to determine during peptide synthesis.

Conjugating Peptide Immunogens to Carrier Protein

Peptides can be conjugated to a carrier protein by several methods. The methods outlined in the following use EDC (1-ethyl-3-[3-dimethylaminopropyl]carbodiimide hydrochloride) and sulfo-SMCC (sulfosuccinimidyl-4-(*N*-maleimidomethyl)cyclohexane-1-carboxylate) cross-linkers.

EDC is a zero-length cross-linker that allows for carboxyls to be covalently attached to primary amines. Depending on the method of coupling, the peptide will conjugate to the KLH in a number of different orientations. This will lead to the generation of antibodies against the peptide sequence presented to the immune system. EDC is suitable for conjugating peptides, which have no terminal cysteine or where a specific peptide orientation is not required.

Sulfo-SMCC is a heterobifunctional cross-linker that possesses both an amine reactive NHS (*N*-hydroxysuccinimide) ester (amino group directed) and sulfhydryl reactive maleimide group (thiol directed), which are connected at opposite ends by a cyclohexane spacer arm. A peptide with a terminal cysteine will conjugate to the KLH with high specificity allowing for a desired epitope (located away from the terminal cysteine) to be presented favourably to the immune system. Sulfo-SMCC is suitable for conjugating peptides that have a terminal cysteine and where specific peptide orientation is desired. A downside of sulfo-SMCC is that antibodies may be raised against the linker.

To prepare an immunogen, the KLH is first reacted with sulfo-SMCC, as the abundance of lysine residues will react with the NHS–ester linker end to form stable amide bonds at pH 7–9. As the conjugation is performed via the primary amines, buffers containing Tris and glycine should be avoided because they will compete [9]. Any excess cross-linker should be separated from the conjugated KLH by gel permeation chromatography. To complete the immunogen, a fully reduced sulfhydryl-containing peptide will react with the maleimide linker end to form stable thioether bonds at pH 6.5–7.5 [8].

Glutaraldehyde is an amine to amine homobifunctional cross-linker as it contains two aldehyde groups, which react with both surface lysines on the carrier protein and amino groups on the N terminus of peptides. The coupling is not as controlled compared with that of heterobifunctional linkers and there is a greater chance of polymerization (as with EDC). The reaction is dependent on pH, temperature and ionic strength [8]. Glutaraldehyde is best suited to coupling peptides that do not contain internal lysine residues or where a terminal lysine is preferred over that of a cysteine. If a peptide contains internal lysine residues, then interpeptide links are formed that can cause the formation of large multimeric complexes and lead to the peptide not being presented properly. As the conjugation is performed via the primary amines, buffers containing tris and glycine should be avoided because they will compete [9].

Once conjugated to an appropriate carrier molecule, the antigen is now ready for immunizing into a suitable animal model.

ANTIBODY PRODUCTION

Once the antigen has been identified, sourced and prepared for immunization (commonly in the form of a synthesized peptide), it is administered into the host animal via injection. Routes of administration include subcutaneous and intradermal. The presence of the foreign molecule stimulates the host's immune response, whereby B cells undergo proliferation into plasma and memory B cells. This process usually takes several days post-immunization and is described as the lag phase. Initially, plasma B cells synthesize IgM subclass antibody. IgM population declines as a result of the T-cell-dependent class switch of IgM producing plasma cells to produce IgG. The first exposure to an antigen is described as the primary response, and as shown in Figure 1.2, it can take several days to generate sufficient antibody to be effective against the antigen. The memory B cells produced during primary exposure enable a quicker response to a subsequent exposure to the same antigen, termed the secondary response. This is the fundamental principle behind vaccinations and the method used to produce large quantities of antibody for commercial use.

Polyclonal Antibody Production

Commercial polyclonal antibodies cover a diverse range of proteins from neuroscience to cancer and metabolism targets. The reason for this is because the polyclonal production

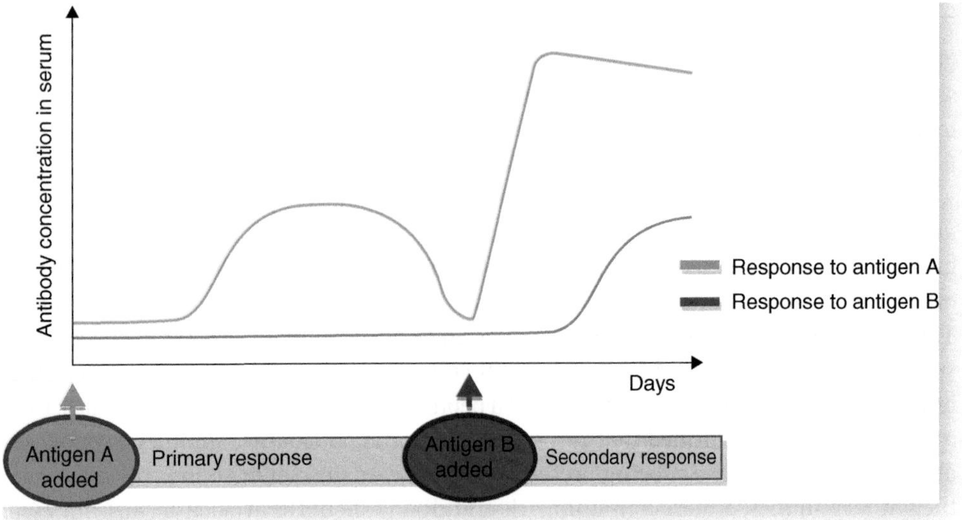

FIGURE 1.2 Typical Primary and Secondary Responses to an Antigen
Upon first exposure to an antigen, a series of B-cell divisions must occur before enough antibody is present to be effective. This lag phase is significantly reduced upon subsequent exposure to the same antigen. Memory B cells generated from the primary response are available to rapidly divide, thus reducing the time required to produce antibodies and generating larger quantities.

process is not complex and becomes cost-effective when generating large quantities of antibody. Unfortunately, the main limitation of polyclonal antibodies is the batch-to-batch variation introduced by differences in the host's immune response. For a particular batch demonstrating specificity with little or no background staining, it is recommended to purchase in bulk wherever possible to ensure completion of a project without introducing ambiguity. Background staining can be reduced by selecting antibodies purified by antigen affinity (see the 'Affinity Purification' Section, p 17).

Mammalian and avian host species are frequently used to generate polyclonal antibodies. Animal selection for polyclonal production should be based on two main criteria: the amount of antibody needed and the requirement for the host and donor species to be phylogenetically different. The latter is a key consideration when selecting secondary detection reagents in order to prevent cross-reactivity with the endogenous immunoglobulins of the host species. The docile nature, low maintenance and robust immune system capable of handling a range of antigen types make rabbits the most widely used mammalian species. Commonly, the inbred New Zealand white rabbit is used, which is a large-sized breed. Larger animals such as donkey, goat and sheep are used where large quantities of antibodies are required (since they produce significantly more serum per specimen than rabbits) or alternative species for use as secondary detection reagents are needed. Polyclonal antibodies can also be raised in rat, mice and guinea pigs. Being small in size, however, means that antibody yield is low compared to their larger mammalian counterparts.

An alternative for large quantities of antibody is the avian species including chicken. Chickens are favoured because their protein repertoire is diverse from mammalians (they are phylogenetically very different), making them particularly useful in the generation of antibodies to a protein target highly conserved between mammalian species. Although a small amount of antibody can be successfully purified from chicken sera, the majority of antibodies are concentrated in the egg yolk. Hence, chicken antibodies are referred to as IgY and form the avian equivalent to mammalian IgG. The structure of chicken IgY differs slightly from IgG, whereby it consists of four constant domains and a single variable region. Approximately 1 g of IgY can be obtained from 12 eggs, which is equivalent to approximately the amount of IgG present in 100 mL of mammalian sera. Harvesting antibodies from egg yolk provides the ethical advantage that invasive bleed procedures associated with mammalian antibody production are eliminated. It is also important to note that egg yolk is a complex mixture of proteins, water and lipids. A delipidation method is required to separate IgY from the lipids for further antigen-affinity purification.

In the process of raising an antibody with the desired affinity and specificity towards a certain protein, it is not uncommon for several animals to be initiated at once. However, it is important to remember the ethical and legal principle of using only the minimum number of individual animals required for the yield of antibody required for a particular project. Between two and six rabbits are typically initiated for polyclonal antibody production. If larger animals are used such as chickens or goats, then one or two is more appropriate. When developing or reviewing a supplier's immunization protocol, key factors to consider are the type of adjuvant, the route of administration and the frequency of immunization and bleeding. Adjuvants are used to induce a prolonged immune response with the aim of increasing antibody titre. Freund's adjuvant is the most commonly used adjuvant, which is of two types: Freund's complete adjuvant and Freund's incomplete adjuvant. The type of adjuvant used may dictate the route of injection due to their potential to have an inflammatory effect on surrounding tissue. Protocol length is usually determined by the speed at which the end user requires the antibody. Suppliers offer short protocols for end users with limited time; however, the volume of sera obtained will be reduced. Longer protocols are more frequently employed to allow time for antibody maturation. Bleeds collected during the early stages of a longer immunization protocol can be analysed to determine relative successfulness of the reagent being produced. A pre-screen can also identify animals generating antibodies of the desired characteristics, which could be maintained on an extended protocol with boost injections to generate large volumes of sera to complete a particular project.

A vast range of custom animal facilities are available for antibody production, many of which have standard immunization protocols in place that can be accessed from their homepage. Many suppliers offer consultation on aspects of the production process, carrier protein conjugation and length of protocol for example, but the elite suppliers will produce an optimized immunization protocol based on the end-user's particular requirements. A key factor to take into account when selecting a supplier is to ensure that their ethical use of animals for research purposes as well as animal welfare procedures conform to the country's required regulations. Legislation for import and export governing the distribution of animal by-products between countries must also be checked when selecting an animal facility in a different country or when intending to distribute the resulting antibodies for use internationally. The United States Department of Agriculture (USDA), for example, provides policies and regulations for the import and export of live animals and animal by-products for the United States.

Monoclonal Antibody Production

Production of monoclonal antibodies follows a more complex process than polyclonal antibody production. Host species for monoclonal antibodies are typically mice and rats. The immune system of these species does not respond efficiently to small antigen molecules; hence, full-length or partial-length recombinant protein immunogens are frequently used. The animal is immunized with the desired antigen to activate B-cell production to a single clone against the antigen. Plasma cells are removed from the animal's spleen and fused with cancerous cells (myelomas) to produce immortal hybridomas that secrete antibodies of the given clone. Screening methods like enzyme-linked immunosorbent assay (ELISA) are used to identify hybridoma colonies with the greatest affinity to the target antigen. These cells are cultured in media until the desired yield is achieved before they are harvested for purification. Monoclonal antibodies are identified by the target name and clone number, for example anti-CD34 clone MEC 14.7. Commercial suppliers occasionally alter the clone number for hybridomas that they license and are usually able to provide additional information upon request.

Increasing in popularity and availability for research and clinical use are rabbit monoclonal antibodies. The detailed technology for generating these antibodies is proprietary; however, the basic principle of hybridoma creation in monoclonal production is applied. Rabbit monoclonals combine the specificity of a monoclonal antibody with the superior antigen recognition system of the rabbit immune response. The result is the generation of a high affinity and consistent reagent. The robust immune system of the rabbit not only provides a platform to employ small immunogens but also generates antibodies capable of functioning across a range of applications. A disadvantage of mouse or rat monoclonal antibodies for immunohistochemistry arises when testing against tissue of the respective species. The secondary antibody is unable to distinguish between antibody and endogenous IgG, which increases background and may obscure specific staining, making interpretation of results difficult, if not impossible. The use of a primary antibody from an alternative host species (e.g. rabbit) is advantageous in these circumstances. Alternatively, a commercially available 'mouse-on-mouse' kit may also be employed.

Monoclonal and Polyclonal Antibodies for Immunochemistry

For immunochemistry, antibodies fall into two basic categories of primary and secondary antibodies.

- Primary antibodies are derived specifically from an antigen (immunogen) in the immunized host. Each specific antibody is produced by a different clone of B cells generating multiple antibody variants, resulting in a population of polyclonal antibodies, each subpopulation recognizing a particular epitope on any given antigen. The serum can then be purified to capture and separate the antibody pool or isolate a specific antibody from the pool. Primary antibodies can also be generated from hybridoma cell lines. Hybridoma cells are antibody producing B cells, which have been fused with an immortal cell line. The advantage of hybridoma cells is that they can be cultured and expanded as individual clones, which secrete only one antibody variant, a monoclonal antibody. The cell culture supernatant can then be clarified (removal of hybridoma cells) and the antibody purified. Primary antibodies recognize and bind the epitope of the antigen they have been raised against with high affinity and specificity.
- Secondary antibodies bind to primary antibodies for the purpose of signal amplification and are typically polyclonal in order to increase the number of secondary antibodies that bind to a single primary antibody. Secondary antibodies can bind specifically to whole immunoglobulin or antibody fragments and are chosen based on the host species and the immunoglobulin subclass of the primary antibody [10]. They are commonly conjugated to a suitable reporter label, usually enzymatic or fluorescent in nature, which is used to visualize the binding location of the primary antibody (see p 25)

Polyclonal antibodies are available as concentrated sera from the immunized animal or, in the case of monoclonal antibodies, as tissue culture supernatant or ascites. However, both are contemporarily supplied in a purified form (purification methods are considered in the 'Fragment Antibody Preparations' Section).

ANTIBODY PURIFICATION

Although crude sera and hybridoma cell culture supernatants can be successfully applied to a number of immunological applications, removal or reduction of antibodies of unwanted specificity (including other irrelevant proteins) can be advantageous to improve the signal relative to background staining. There is an availability of a huge range of commercially available resins and a number of techniques, which can be employed for purifying whole IgG or fragment preparations from sera and supernatants. In the following, the summary of each of the techniques most commonly applied in the purification of monoclonal and polyclonal antibodies from sera and cell culture supernatants is provided.

Ammonium Sulphate Precipitation

A quick and simple process for the removal of protein impurities from sera and supernatants is ammonium sulphate precipitation. It is well documented in the purification of monoclonal antibodies where high recoveries and a significant reduction in impurities can be achieved [11]. The solubility of antibodies varies according to the ionic strength of the solution. Ammonium sulphate competes with the antibodies for hydrogen bonding with the water molecules, so increasing the ionic strength of the solution by the addition of ammonium sulphate causes the antibodies to precipitate out. The antibodies can then be recovered by centrifugation, leaving the majority of protein impurities in solution. Although ammonium sulphate is a crude but effective purification method, it will generally not deliver antibodies to the same level of purity achieved through affinity purification techniques. Also, the harsh nature of ammonium sulphate precipitation has a deleterious effect on the specificity of some antibodies.

Affinity Purification

All biological processes depend on specific interactions between molecules. Affinity chromatography applies this principle in the case of antibody affinities for adsorption to a solid phase. A ligand (commonly the immunogen used to raise the antibody) is immobilized on the solid phase (matrix) and the counter-ligand (in this case the antibody) is adsorbed from the sera or cell culture supernatant that is passed down the column.

Ligands used for affinity chromatography applications typically possess the following characteristics:

- Be highly specific towards the target to be isolated (in this case, a specific subset of antibodies).
- Possess chemical properties that allow for coupling to a chromatography matrix.

- Be able to form reversible complexes to allow for the antibody to be isolated (bind) and then separated (eluted) from the matrix through a subtle change in conditions (e.g. buffer pH or ionic strength).

There are a vast array of ligand types that can be employed (mono-specific and group-specific), which fall outside the scope of this book.

A basic process outline of affinity purification (outlined diagrammatically in Fig. 1.3):

1. The ligand-bound affinity column is equilibrated in binding buffer to create conditions that favour specific binding of the antibody to the ligand.
2. The sera or cell culture supernatant is next applied to the column. Antibody binds specifically (but reversibly) to the column ligand and any unbound material washes through the column.
3. The antibody is recovered from the column by changing conditions to favour elution (disrupt the antibody–antigen complex).
4. The affinity column is re-equilibrated and cleaned to remove any traces of antibody (or material bound to the matrix) not eluted in step 3.

Examples of commercially available affinity matrices are summarized in the following [12].

Proteins A and G Resins

Proteins A and G are used widely in the purification of monoclonal cell culture supernatants. Both proteins A and G media are manufactured from bacteria-derived proteins

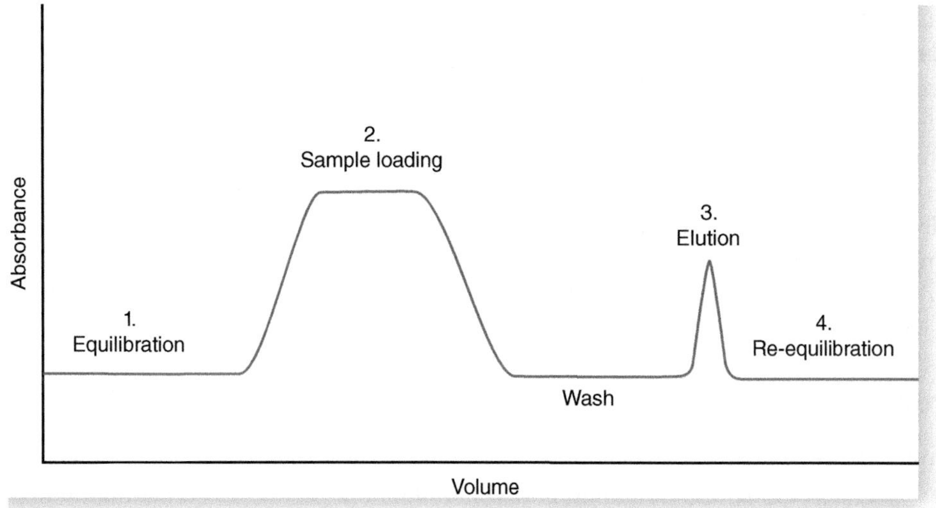

FIGURE 1.3 An Outline Affinity Purification Chromatogram

and immobilized to matrices for affinity purification. Protein A is a bacterial protein from Group A *Staphylococcus aureus*, and protein G is a bacterial protein from Group G *Streptococci aureus*. Both proteins A and G bind with the Fc region of immunoglobulins through interaction with the heavy chain (CH_3–CH_2 hinge region). In addition to the Fc region, protein G will also bind to the CH_1 region [13]. Proteins A and G resins display different relative binding strengths for different immunoglobulins.

Protein A has a large number of beneficial characteristics, which make it a popular choice for both large- and small-scale platforms. The benefits include but are not exhausted to:

- It is stable across a wide pH range (2–11).
- It delivers high purity of 95%to $\geq$ 99% [11].
- It can refold after being exposed to a range of denaturing agents.

Although sensitive to alkaline conditions (recombinant variants have greater alkali stability), the lack of cysteine residues allows for effective cleaning with reducing agents [11]. Protein A can bind all subclasses of mouse IgG including IgG1, which is often documented as having a low affinity to protein A but can be efficiently purified using certain buffer conditions [12].

Protein A is a polypeptide which as a full-length molecule has a cell wall membrane (C terminus), linking five homologous antibody-binding domains (E, D, A, B and C). Each antibody-binding domain is ~6.6k Da in size with a full-size protein A molecule being ~54 kDa. However, when coupled with a matrix in commercially prepared media the cell wall domain is often deleted, reducing the molecule to a size of ~42 kDa [11].

Protein G is a type III Fc-receptor cell surface protein and can be used to purify a broader range of IgG classes than protein A. Protein G can capture all mouse and human subclasses as well as rat IgG_{2a} and IgG_{2b}. Protein G generally has a greater affinity for IgG than protein A. However, a lower elution pH (2.5–3) may be required depending on the antibody. The lower elution pH may impact the biological activity of the antibody being purified, so protein A may be preferable as it generally employs milder elution conditions [10].

Protein G has two immunoglobulin-binding sites as well as cell surface and albumin-binding sites. Commercially prepared media typically use recombinant variants that have had both the cell surface and albumin-binding regions genetically deleted to reduce non-specific binding. The native protein G molecule is ~65 kDa with the recombinant protein ~21.6 kDa in size [10].

Protein L

Protein L originates from a bacteria-derived protein *Peptostreptococcus magnus*, much like proteins A and G. However, protein L differs from both proteins A and G in that it binds antibodies via the V_L region (variable region of light chain) of the Fv part (fragment of the variable regions). Since protein L does not interact with antibody heavy chains, it is able to bind a greater number of immunoglobulin classes than either protein A or G. Protein L is therefore able to bind IgG, IgM, IgA, IgE, IgD as well as Fab fragments (fragment antigen-binding region), scFv (single-chain variable fragment) and sdAbs (single-domain antibodies). Protein L is restricted to only binding antibodies that have kappa light chains and further limited to specific subtypes. Due to the specific binding characteristics of protein L, it is recommended to be used only for the purification of monoclonal antibodies that possess kappa light chains. The main advantage to the use of protein L compared with proteins A and G for the purification of monoclonal antibodies from cell culture supernatant is that protein L will not bind bovine immunoglobulin if used as a culture media supplement.

The molecular weight of recombinant protein L used for the preparation of commercial resins is ~36 kDa. The binding affinity of proteins L, A and G for antibodies is summarized in Table 1.2.

Ion Exchange

Avian immunoglobulins (IgY) prepared from the immunization of hen's eggs do not bind to either protein A, G or L resin, so an alternative purification method is required. Ion-exchange chromatography (IEC) purifies antibodies and other proteins through ionic interactions and has been used since the late 1940s [12]. IEC can be successfully applied to the purification of IgY as it separates proteins by charge; thus, it is not restricted by the binding characteristics of an affinity ligand.

NHS and Iodoacetyl-Activated Resins

N-Hydroxysuccinimide (NHS) and iodoacetyl-activated media are used widely in the purification of polyclonal antibodies from sera. The immunizing peptide is immobilized onto a matrix (typically beaded agarose) as the ligand for purifying antibodies raised specifically against it. There are a wide range of commercially pre-activated resins available, which fall outside the scope of this chapter [12].

TABLE 1.2 Binding Affinity of Antibodies for Proteins L, A and G

Species	Immunoglobulin	Protein L	Protein A	Protein G
Mouse	IgG1	Strong binding	Weak binding	Strong binding
	IgG2a	Strong binding	Strong binding	Strong binding
	IgG2b	Strong binding	Strong binding	Strong binding
	IgG3	Strong binding	Weak binding	Strong binding
	IgM	Strong binding	Weak binding	No binding
Rat	IgG	Strong binding	Weak binding	Strong binding
Rabbit	IgG	Weak binding	Strong binding	Very Strong binding
Goat	IgG	No binding	Weak binding	Strong binding
Chicken	IgY	Strong binding	No binding	No binding

NHS media covalently couples with the immunizing peptide via a primary amine to form an amide bond. NHS media can be used for coupling peptides that do not have a terminal cysteine or possess an internal cysteine (where coupling to the matrix would be unfavourable). Since the coupling is via a primary amine, the peptide will be coupled with the matrix in multiple orientations; therefore, the desired epitope may not be best positioned for optimal antibody binding. Iodoacetyl media, however, couples with the immunizing peptide specifically via a free sulfhydryl (side chain in cysteine) to form a stable thioether bond. A peptide with a terminal cysteine will couple with the matrix in a single orientation allowing for a desired epitope (located away from the terminal cysteine) to be presented favourably for optimal antibody binding.

FRAGMENT ANTIBODY PREPARATIONS

Fab and $F(ab)_2$ antibody fragments can offer a number of advantages over the use of whole IgG in immunochemical studies. Antibody fragments can offer greater penetration into tissues and do not suffer from non-specific adsorption by Fc receptors. A downside to the use of Fab fragments is loss of avidity [9]. However, Fab fragments are extremely effective for blocking endogenous immunoglobulins on cells and tissues when using primary antibodies raised in the same species as the test tissue. Pepsin and papain are enzymes typically used in the digestion of whole IgG into Fab, $F(ab)_2$ and Fc fragments. Immunoglobulins sourced from different classes differ in their sensitivity towards proteolytic digestion.

$F(ab)_2$ fragments are prepared by performing a pepsin digest. The optimum acidic conditions for pepsin function is ~pH 2; however, at such an acidic pH, irreversible damage is likely to occur to the antibody and so conditions no lower than pH 4 are recommended. Typically, the antibody and pepsin are incubated at 37 °C in sodium acetate buffer (pH 4–4.5) for a predetermined period of time (Ig class specific). The digest is quenched by neutralizing the acidic conditions using 2 M Tris. An alternative to the use of free pepsin is to use a commercially available immobilized pepsin resin. The $F(ab)_2$ digest can then be separated from any undigested IgG and Fc fragments by protein A affinity chromatography. Only the intact IgG and Fc fragments will bind to the matrix allowing for the $F(ab)_2$ fragments to be collected in the flow-through fraction(s) [9].

Fab fragments are prepared by performing a papain digest. Typically, the antibody and papain are incubated in the presence of a reducing agent (cysteine) for a predetermined period of time (Ig class specific). The digest is stopped using an oxidant (iodoacetamide). However, similar to pepsin, an alternative to the use of free papain is to use a commercially available immobilized papain resin. The Fab digest can then be separated from any undigested IgG and Fc fragments by protein A affinity chromatography. Only the intact IgG and Fc fragments will bind to the matrix allowing for the Fab fragments to be collected in the flow-through fraction(s). Fragmentation of IgG by papain and pepsin is diagrammatically shown in Figure 1.4.

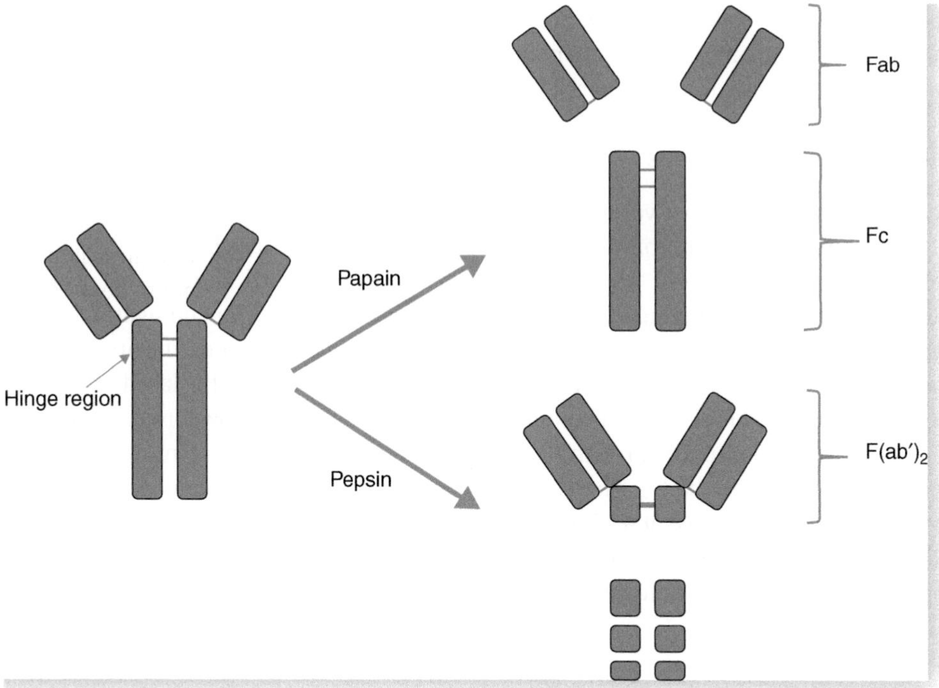

FIGURE 1.4 Fragmentation of IgG Immunoglobulin by Papain and Pepsin

If non-immobilized papain and pepsin are used to increase the purity of the final fragment pool and remove any remaining enzyme, the Fab (~50 kDa) and $F(ab)_2$ (~110 kDa) preparations can be loaded onto a size-exclusion column and fractions are collected corresponding to the Fab and $F(ab)_2$ fragment molecular weights.

ANTIBODY LABELLING

Reporter molecules used in immunochemistry are dependent on the detection method being employed. The two most popular detection methods are enzymatic and fluorescence using either an antibody conjugated to an enzyme or fluorophore. In chromogenic detection methods, an enzyme label reacts with a suitable substrate in order to produce a visual precipitate at the site of antibody binding that can be observed with a traditional light microscope. Horseradish peroxidase (HRP; ~44 kDa) and calf intestinal alkaline phosphatase (AP; ~140 kDa) are the two most commercially used. Fluorescence detection methods typically use a small chemical compound (fluorophore) that is excited at a specific wavelength and emits light at a longer wavelength. Historically, commonly used fluorophores in immunochemistry are fluorescein isothiocyanate (FITC), tetramethylrhodamine-5-(and 6)-isothiocyanate (TRITC) and cyanine. More recently, contemporary (so-called 'second-generation') fluorophores, such as Alexa Fluor®, offer increased photostability and brighter staining (have higher quantum yields) when compared to their historical counterparts. For more information on reporter labels, see p 25.

Fluorescent Reporters

FITC is an amine-reactive variant of fluorescein dye incorporating an isothiocyanate-reactive group to facilitate conjugation to antibodies via lysine side chains. FITC is readily soluble in aqueous buffers, and conjugation can be performed between pH 7 and 9 (Fig. 1.5). However, coupling efficiency is increased at >pH 8. The conjugation can be performed at room temperature with the incubation time dependent on the antibody and the degree of labelling that is required. It is recommended to separate the labelled antibody from any unreacted FITC using a desalting column or dialysis to allow for accurate determination of the fluorophore:protein (F:P) ratio (the number of fluorescent dye molecules coupled with each antibody). The F:P of the antibody/FITC conjugate is determined by measuring the absorbance at 495 nm (FITC) and 280 nm (protein, in this case antibody) [9]. Alternative pre-activated variants of fluorescein dye include fluorescein-5-maleimide (F5M), which allows for conjugation via a free sulfhydryl.

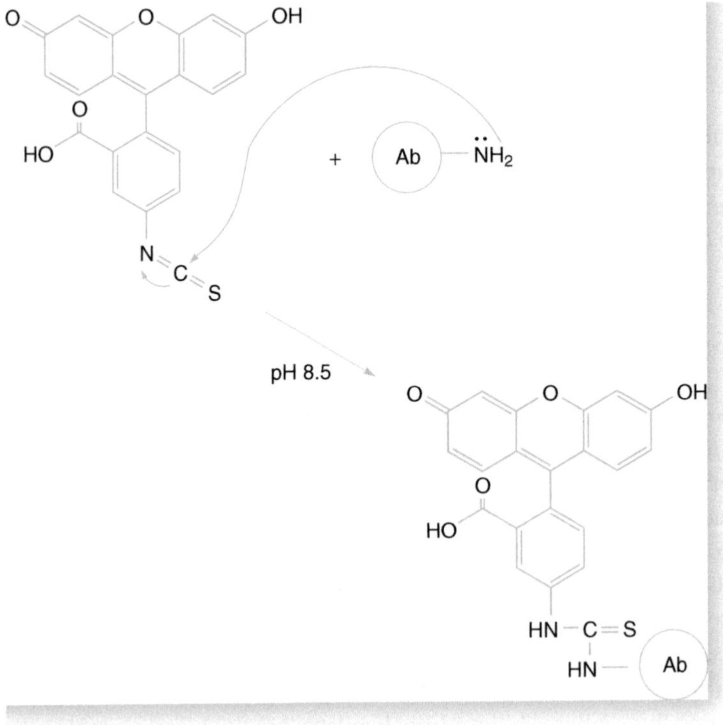

FIGURE 1.5 Fluorescent Labelling of Antibody with FITC

TRITC is an amine-reactive variant of rhodamine dye incorporating an isothiocyanate-reactive group to facilitate conjugation to antibodies via lysine chains. TRITC is soluble in aqueous buffers, and the coupling conditions used for FITC are also applicable to the efficient labelling of antibodies to TRITC. The F:P of the antibody/TRITC conjugate is determined by measuring the absorbance at 596 nm (TRITC) and 280 nm (protein, in this case antibody).

Enzyme Reporters

There are a number of different (multi-stage) methods that can be followed for the successful conjugation of enzyme reporter labels to antibody, which fall outside the scope of this book. An overview of a recommended protocol used for the conjugation of HRP to antibodies suitable for biological model experiments is detailed in the following.

The conjugation is performed over a three-step process:

1. *Sulfo-SMCC HRP activation*
   The HRP is first reacted with sulfo-SMCC (heterobifunctional cross-linker), coupling the linker with the HRP via the amine-reactive NHS ester. The reaction is incubated at room temperature, and sufficient coupling is achieved after ~50 min. Any excess sulfo-SMCC is removed using a desalting column.
2. *2-Iminothiolane (2-IT or Traut's reagent) antibody thiolation*
   2-IT is a thiolation compound that reacts with a primary amine to introduce a sulfhydryl group presented at the end of a short spacer arm. The 2-IT is incubated with the antibody at room temperature for ~12 min, and the reaction stopped with the addition of 1 M glycine. Any excess 2-IT is removed using a desalting column.
3. *Conjugate-activated HRP and thiolated antibody*
   The activated HRP and thiolated antibody are reacted together at a defined ratio and incubated at room temperature for ~2–4 h. Any excess HRP is removed using a size-exclusion column and the degree of labelling determined (the number of HRP molecules coupled with each antibody).

ANTIBODY STABILITY AND STORAGE

Antibodies are generally stable. However, optimal conditions for storage are unique to each antibody. Some general guidelines for the storage of purified antibodies can be applied to help prevent aggregation and retain biological activity. Increased background staining in immunochemical studies may be attributed to the presence of aggregates due to hydrophobic interactions.

For short-term storage, antibodies can be stored at 2–8 °C where the use of a preservative such as sodium azide (0.002–0.05%) is recommended to help prevent bacterial or fungal growth. If antibodies are to be used for *in vivo* experiments, sodium azide should be avoided. As an alternative, it is recommended to sterile filter to prolong storage time.

For long-term storage, most antibodies are best stored at −20 to −80 °C. In addition to the use of a preservative, a cryoprotectant (such as glycerol) can be used to help reduce the effects of freeze–thaw events by preventing the antibody solution from freezing. Antibodies are sensitive to repeated freeze–thaw cycles (decrease biological activity) and should be aliquoted where appropriate and/or a cryoprotectant used. If antibodies are to be stored at concentrations of <1 mg/mL, the addition of a bulk protein (stabilizing agent) can be used to improve stability. BSA is typically used at 0.1–1% (w/v). It should be noted that the addition of stabilizing/preservative agents will reduce the purity of the antibody preparation and may need to be removed at a later date [10]. Antibodies that are contaminated by microbial or bacterial growth should not be used in biological assays, as in immunochemical studies, high background levels and/or poor staining may be observed.

It should be noted that sodium azide is not recommended for preserving antibodies that are intended for conjugation to an enzyme or fluorochrome, as in concentrations >3 mM, it may interfere with antibody labelling (reduce coupling efficiency). Sodium azide should not be used as a preservative for HRP-conjugated antibodies as sodium azide inhibits HRP. An alternative to the use of sodium azide when preparing and storing antibody conjugates is borate-buffered saline. It can be used to help prevent microbial and bacterial growth.

REFERENCES

1. Charles, A.J., Travers, P., Walport, M. and Shlomchik, M.J. (2001) *Immunobiology the Immune System in Health and Disease*, 5th edn, Garland Publishing, New York, NY, USA.
2. Matthews, C.K., van Holde, K.E. and Ahern, K.G. (1999) *Biochemistry*, 3rd edn, Benjamin/Cummings, San Francisco, CA, USA.
3. Larsen, J.E., Lund, O. and Nielsen, M. (2006) *Immunome Research*, **2**, 2.
4. Saper, C.B. (2009) *Journal of Histochemistry & Cytochemistry*, **57**, 1–5.
5. Hopp, T.P. and Woods, K.R. (1981) *Proceedings of the National Academy of Sciences of the United States of America*, **78**, 3824–3828.
6. Soga, S., Kuroda, D., Shirai, H. *et al.* (2010) *Protein Engineering, Design & Selection*, **23**, 441–448.
7. Kuroda, D., Shirai, H., Jacobson, M.P. and Nakamura, H. (2012) *Protein Engineering, Design & Selection*, **25**, 507–521.
8. Wong, S.S. and Jameson, D.M. (2012) *Chemistry of Protein and Nucleic Acid Cross-Linking and Conjugation*, 2nd edn, CRC Press, Florida, USA.
9. Ahmed, H. (2005) *Principles and Reactions of Protein Extraction, Purification and Characterization*, CRC Press LLC, Florida, USA.
10. GE Healthcare (2002) *Antibody Purification Handbook*, GE Healthcare, Uppsala, Sweden.
11. Gottschalk, U. (2009) *Process Scale Purification of Antibodies*, John Wiley & Sons Inc., Hoboken, NJ, USA.
12. Janson, J.-C. and Ryden, L. (1998) *Protein Purification*, 2nd edn, John Wiley & Sons Inc., New York, NY, USA.
13. Hagel, L., Jagschies, G. and Sofer, G. (2008) *Handbook of Process Chromatography*, 2nd edn, Academic Press, London, UK.

CHAPTER TWO

The Selection of Reporter Labels

Judith Langenick

AbD Serotec, Oxford, UK

INTRODUCTION

In an immunochemical experiment, a primary antibody specifically binds to a certain protein (see p 35). This binding is visualized microscopically with the help of a detection (signal amplification) system, which provides information about the presence and location of the respective protein. Since the first IHC (immunohistochemical) experiments by Dr Albert Coons in 1941, much progress has been made to increase the number and efficiency of detection systems. In the beginning, secondary antibodies conjugated to enzymatic reporter labels were mainly used for visualization (immunoenzymatic detection), which resulted in coloured precipitate forming at the site of primary antibody binding when exposed to their substrates. Although enzymatic reporter labels are still routinely used, directly conjugated primary antibodies and fluorescent reporter labels have recently increased in popularity. The latter is mainly due to the availability of fluorescent dyes in a great variety of colours (with defined excitation and emission profiles) and the

Immunohistochemistry and Immunocytochemistry: Essential Methods, Second Edition. Edited by Simon Renshaw.

development of a new generation of bright and photostable fluorescent dyes. The demand for these dyes mainly emerged from the need to visualize multiple antigens in the same specimen.

Compared to serial sections, multicolour immunostaining does not only save time and reagents but also provide a solution for small cytological samples for which serial sections are highly problematic. Multiple immunostaining also facilitates the detection of two or more proteins that are co-localized in a single subcellular location. When two proteins are located in different cellular compartments, each protein is detected in a single colour. However, if the proteins co-localize, a merged colour (the result of both individual colours) is visible. Such experiments have helped to further increase the understanding of protein–protein interactions. This chapter will focus on the basic principles of enzymatic and fluorescent reporter labels.

ENZYMATIC LABELS

An immunoenzymatic staining reaction of an enzyme with its substrate results in the change of a chemical compound (chromogen) to a coloured precipitate. This process can be summarized in the following formula [1]:

1. Enzyme (E) + Substrate (S) = ES complex (rather transient)
2. $ES \rightarrow E +$ Product (P)

When using an immunoenzymatic detection method, the enzyme tends to be covalently or non-covalently bound to a suitable signal amplification system. These include avidin–biotin complex (ABC) and primary or secondary antibodies. In general, these will be purchased from a commercial company. The use of directly conjugated primary antibodies in immunochemistry is possible, but only recommended for highly abundant antigens. This is due to the fact that the extra signal amplification provided by the binding of several secondary antibodies carrying multiple enzymes to a primary antibody is missing. The lack of this amplification step can result in weak or absent staining. After the antibody incubation step, a chromogenic substrate is added to react with the enzymes present on the antibodies or ABC complex. Upon contact with the enzyme/substrate complex, the chromogen converts into an insoluble product that precipitates at the site of the detection system, which corresponds to the primary antibody-binding site.

The most commonly used enzymatic IHC labels are calf-intestinal alkaline phosphatase (AP) and horseradish peroxidase (HRP). Both are stable in solution, relatively cheap and readily available in large quantities. They also lead to fairly stable and easily microscopically detectable products. The enzyme and chromogenic substrate combination determines the final permanent staining colour. For example, adding DAB (3,3′ diaminobenzidine) chromogen to HRP results in a brown precipitate while the addition of AEC (3-amino-9-ethylcarbazole) chromogen leads to red. The most popular chromogens for both HRP and AP and the respective chemical reactions to produce the coloured precipitates are described in the following.

HRP

Upon the addition of a chromogenic substrate, an enzyme–substrate complex is formed. This step is followed by the oxidation of the chromogen. Hydrogen peroxide is the oxidizing agent for this reaction, with HRP catalysing the transfer of two electrons from the chromogen to hydrogen peroxide, with the chromogen therefore becoming oxidized. Water is formed as a by-product. Once the chromogen is fully oxidized, the catalysis of hydrogen peroxide is terminated and the reaction is stopped. An example of a typical HRP reaction (in this instance using the chromogen DAB) is as follows:

$$\mathrm{HRP} + 2\mathrm{H_2O_2} + \mathrm{DAB} \rightarrow \mathrm{HRP} + 2\mathrm{H_2O} + \mathrm{O_2} + \text{oxidized DAB}$$

(brown-coloured end product)

Table 2.1 gives an overview of common HRP chromogens.

Considerations When Using HRP

Many tissue sections contain red blood cells, which are rich in endogenous peroxidases that will react with chromogen, resulting in false-positive staining. It is therefore recommended

TABLE 2.1 Overview of HRP Chromogens

Chromogen Name	Colour of End Product	Comments
3,3′-Diaminobenzidine (DAB)[a]	Brown	Insoluble in alcohol and other organic solvents. Dehydration before coverslipping is therefore possible, allowing for the use of organic mounting media, such as DPX (see p 75)
3-Amino-9-ethylcarbazole (AEC)[a]	Red	Due to its high solubility in alcohol, aqueous counterstains and mounting media have to be used (see p 75). Store specimens in the dark to avoid fading caused by light exposure
4-Chloro-1-naphtol (CN)	Blue	Due to high solubility in alcohol and other organic solvents, aqueous counterstains and mounting media have to be used (see p 75). Has a tendency to diffuse away from the site of initial precipitation during storage
p-Phenylenediamine dihydrochlo-ride/pyrocatechol (Hanker-Yates reagent)	Blue/black	Insoluble in alcohol and other organic solvents. Dehydration before coverslipping is therefore possible, allowing for the use of organic mounting media, such as DPX (see p 75)

[a]Popular choice.

to test samples for endogenous peroxidase activity by adding the substrate of choice in the absence of HRP. If a stained precipitate appears, an alternative enzymatic label such as AP should be considered. Alternatively, endogenous peroxidases can be inhibited with commercial peroxidase suppressors or 3% (v/v) H_2O_2 in methanol, water or a suitable buffer (PBS/TBS) (see p 61) [2].

HRP is inhibited in the presence of bacteriostatic agents such as sodium azide and chemicals such as methanol, sulfides and cyanide, and to the degradation by micro-organisms [3, 4].

AP

AP hydrolyses (and therefore removes) naphthol phosphate groups from its substrates (organic esters, e.g. BCIP), creating phenolic compounds. These compounds then couple with diazonium salts (e.g. INT, NBT, or TNBT), reducing them to form a coloured precipitate (formazan) at the primary antibody-binding site. Important enzyme co-factors for this reaction are Mn^{2+}, Ca^{2+} and Mg^{2+} (Table 2.2).

TABLE 2.2 Overview of AP Chromogens

Substrate	Chromogen	Colour of End Product	Comments
Naphthol AS-MX phosphate	Fast Red TR[a]	Red	Due to their high solubility in alcohol, aqueous counterstains and mounting media have to be used (see p 75).
	Fast Blue BB[a]	Blue	
Naphthol AS-BI phosphate	New Fuchsin[a]	Red	Insoluble in alcohol and other organic solvents. Dehydration before coverslipping is therefore possible, allowing for the use of organic mounting media, such as DPX (see p 75). Brighter than Fast Red TR.
BCIP (5-bromo-4-chloro-3-indolyl phosphate, *p*-toluidine salt)	INT (*p*-iodonitrotetrazolium chloride)	Brown	Due to its high solubility in alcohol, aqueous counterstains and mounting media have to be used (see p 75).
	NBT (nitroblue tetrazolium chloride)	Deep purple	Insoluble in alcohol and other organic solvents. Dehydration before coverslipping is therefore possible, allowing for the use of organic mounting media, such as DPX (see p 75). TNBT produces a deeper purple colour than NBT.
	TNBT (tetranitro blue tetrazolium)	Deep purple	

[a]Popular choice.

Considerations When Using AP

As with endogenous peroxidases, some samples might contain endogenous phosphatases. To test for endogenous phosphatase activity, NBT/BCIP is added to determine if a coloured precipitate forms. If present, endogenous phosphatase activity can be blocked by the addition of the phosphatase inhibitor, Levamisole. Levamisole suppresses endogenous phosphatase activity and therefore reduces background staining, although not in placenta or the small intestine. It can be used between the primary and secondary antibody incubation steps [2] or in the chromogen solution (see p 61). The AP label itself will not be affected due to its intestinal origin.

Glucose Oxidase and Beta-Galactosidase (ß-GAL)

Two other far less common enzymatic labels are glucose oxidase and beta-galactosidase. These labels are often selected for the staining of samples containing high amounts of endogenous peroxidase and/or phosphatase activity. In mammalian tissue samples, no endogenous glucose oxidase or beta-galactosidase are present [3, 5].

The reaction of beta-galactosidase with 5-bromo-4-chloro-3-indolyl-β-D-galactopyranoside (BCIG or X-gal) leads to a bright blue precipitate.

Glucose oxidase catalyses the oxidation of β-D-glucose generating hydrogen peroxide and gluconic acid. Very often, the activity of glucose oxidase is visually shown with the help of tetrazolium salts and intermediate electron carriers. For example, when coupling the tetrazolium salt NBT with the intermediate electron carrier MPMS (1-methoxyphenazine methosulfate), NBT becomes reduced resulting in a purple/blue precipitate [6]. Organic mounting media are recommended.

FLUORESCENCE DETECTION

Fluorescence is a three-stage process involving chemical compounds known as fluorochromes (or fluorophores) (see Fig. 2.1).

Once these compounds are illuminated by a light source, an electron surrounding the fluorochrome's atomic nucleus absorbs energy from photons of light. The result of this is an increase in the energy level of the electron from the ground state (S_0) to the excited singlet state (S_1'). This first step of the fluorescence process is referred to as 'excitation'.

Compared to the highly stable ground state, the S_1' is very unstable and only lasts for a couple of nanoseconds. The short lifetime of this state is due to the fluorochrome undergoing internal conformational changes, termed 'internal conversion'. The results of this change is a loss of some of the electron's energy as heat and the electron falling to a lower, more stable energy level called the relaxed singlet state (S_1). Once the electron moves from the S_1 state back to its ground state, the remaining energy taken up during the excitation process is released as light. This process results in the fluorochrome emitting light at

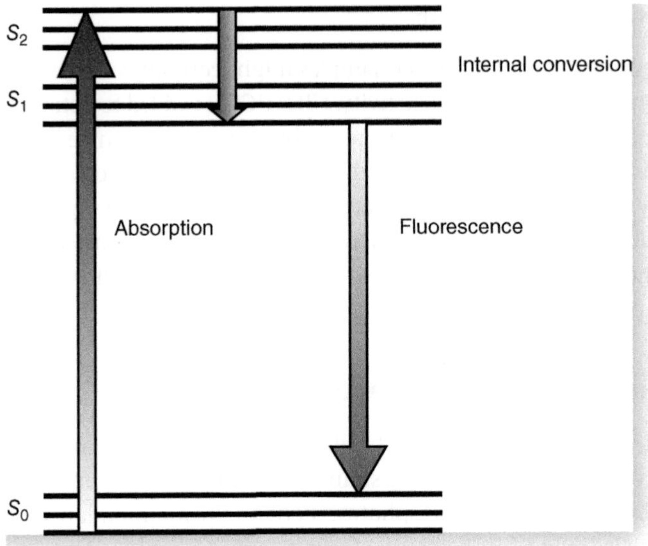

FIGURE 2.1 Diagram of Fluorochrome Excitation. The Addition of Light of a Certain Wavelength to a Fluorochrome Results in the Excitation of an Electron from the S_0 (Ground State) to the Unstable S_1' State. In this State, the Electron only Remains for a Few Nanoseconds before a Conformational Change Results in the Release of Heat and a Drop to the More Stable S_1 State. Once the Electron Moves from the S_1 State Back to S_0, the Remaining Energy Taken up During the Excitation Process is Emitted as Light of a Longer Wavelength than the One Initially Absorbed

a longer wavelength than the one it had absorbed. The wavelength difference between the maximal points of the absorption and emission spectra maxima is known as 'Stoke's shift'.

It should be noted that the excitation/emission cycle can be repeated by a single fluorochrome thousands of times, unless the fluorochrome has been photobleached (see p 32).

Fluorochrome Characteristics

The biggest advantage of using fluorescent reporter labels over enzymatic ones is the sheer variety of bright, photostable dyes available, such as those belonging to the Alexa Fluor® and DyLight Fluor® dye ranges. However, it is important to bear in mind that not every fluorochrome is suitable for every application. For example, while PE (phycoerythrin) is one of the brightest fluorochromes on the market and routinely used in flow cytometry, it is not recommended for use in microscopy due to its quick photobleaching characteristics.

In general, fluorochromes will only optimally fluoresce once they are illuminated with light of a suitable wavelength (depending on the absorption spectrum of the respective fluorochrome and if the delivered energy is sufficient enough to excite the electrons). Also, illumination of a sample with an unsuitable filter set located in the optical path might

TABLE 2.3 Absorption and Emission Spectra for Common Fluorochromes

Label	Maximum Absorption (nm)	Maximum Emission (nm)	Colour
Alexa Fluor® 405	402	421	Blue
AMCA	346	442	Blue
Cy2	489	506	Dark green
DyLight® 488	493	518	Green
Alexa Fluor® 488	495	519	Green
FITC	494	520	Green
Alexa Fluor® 555	555	565	Green/yellow
Cy3	550	570	Green/yellow
Phycoerythrin	488	575	Yellow
TRITC	557	576	Yellow
DyLight® 550	562	576	Yellow
Rhodamine	544	576	Yellow
TAMRA	557	583	Yellow/orange
Cy3.5	576	589	Yellow/orange
Alexa Fluor® 568	578	603	Orange
PE/Texas Red®	535	615	Orange
Texas Red®	595	615	Orange
Alexa Fluor® 594	590	617	Orange/red
DyLight® 594	593	658	Orange/red
APC	650	662	Red
Cy5	643	667	Red
Alexa Fluor® 647	650	668	Red
PE/Cy7	488	670	Red
DyLight® 650	652	672	Red
Cy5.5	675	694	Far red
PE/Cy5.5	535	694	Far red
Alexa Fluor® 680	679	702	Far red
Alexa Fluor® 750	749	775	Far red
APC/Cy7	633	776	Far red
Alexa Fluor® 790	784	814	Far red

lead to little or no fluorescence at all. It is therefore important to check both the excitation and emission spectra of the fluorochrome intended to be used and the microscope filter sets. Information about fluorochrome absorption and emission spectra can be found on the websites of all major antibody companies, imaging core facilities and microscope manufacturers. Table 2.3 contains absorption and emission spectra for common fluorochromes.

When selecting the optimal fluorochrome, the following characteristics should be fulfilled:

1. Spectral properties of the fluorochrome are compatible with the microscopic set-up, and optimal excitation of the fluorochrome is possible.
2. The fluorochrome is bright as indicated by the following:
   a. It has a high extinction coefficient (the probability to absorb a photon of light at a specific wavelength). For example, the bright and commonly used Alexa Fluor® 647 dye has an extinction coefficient of 270,000 compared to 35,000 of the comparatively dim Alexa Fluor® 405 [7].
   b. It has a high quantum yield. This is a measure of the efficiency of the fluorescence process and is calculated by dividing the number of photons emitted by the number of photons absorbed. The maximum quantum yield possible is 1, and bright fluorochromes such as Alexa Fluor® 488 have a high quantum yield of 0.92. However, other Alexa Fluor® dyes such as Alexa Fluor® 555, which has a quantum yield of 0.1, are still regarded as reasonably fluorescent [8].
3. The fluorochrome is highly photostable and has a low susceptibility to photobleaching. First-generation fluorochromes such as FITC are especially susceptible to a self-destructive photochemical process known as photobleaching, which results in quick fading of the fluorescence signal. Although the exact mechanism is still elusive, it is known that it is caused by reactive oxygen species generated during the excitation process. To reduce photobleaching, it is recommended to use next-generation fluorochromes with high photostability such as Alexa Fluor® or DyLight Fluor® dyes. Alternatively, photobleaching can be minimized by using antifade reagents (singlet oxygen scavengers) and by reducing the intensity and the exposure time to the excitation light.
4. The fluorochrome stays fluorescent over a broad pH range.

Some tissues especially from plants and those rich in collagens and lipofuscin tend to autofluoresce. As autofluorescence can overlap with a specific fluorescence, it is advisable to include unlabelled samples in the experimental design to control for this phenomenon [9]. For more information regarding autofluorescence, please see p 62.

REFERENCES

1. Agilent Technologies (2009) Education Guide: Immunohistochemical Staining Methods Edition 5, http://www.dako.com/08002_03aug09_ihc_guidebook_5th_edition_chapter_15.pdf (accessed 15 July 2014)

2. Snider, J. (Thermo Fisher Scientific Inc.) Blocking Endogenous Targets for IHC. http://www.piercenet.com/method/methods-block-endogenous-targets (accessed 15 July 2014)
3. Snider, J. (Thermo Fisher Scientific Inc.) Enzyme Probes. http://www.piercenet.com/method/enzyme-probes (accessed 15 July 2014).
4. Straus, W. (1974) Cleavage of Heme from Horseradish Peroxidase by Methanol with Inhibition of Enzymatic Activity. *Journal of Histochemistry & Cytochemistry*, **22** (908). doi: 10.1177/22.9.908
5. Kuhlmann, W. D. (2010) GOD Cytochemistry. http://www.immunologie-labor.com/cellmarker_files/IET_meth_GOD_cyto.pdf (accessed 15 July 2014).
6. Van der Loos, C. M. User Protocol: Practical Guide to Multiple Staining. Cambridge Research & Instrumentation, Inc. http://www.biotechniques.com/multimedia/archive/00074/CRI-FP-Microscopy_74545a.pdf (accessed 15 July 2014).
7. Thermo Fisher Scientific Inc. The Alexa Fluor Dye Series–Note 1.1. http://www.lifetechnologies.com/uk/en/home/references/molecular-probes-the-handbook/technical-notes-and-product-highlights/the-alexa-fluor-dye-series.html (accessed 15 July 2014).
8. Thermo Fisher Scientific Inc., Fluorescence Quantum Yields (QY) and Lifetimes (τ) for Alexa Fluor Dyes—Table 1.5. https://www.lifetechnologies.com/uk/en/home/references/molecular-probes-the-handbook/tables/fluorescence-quantum-yields-and-lifetimes-for-alexa-fluor-dyes.html (Accessed: July 15, 2014).
9. North, A.J. (2006) Seeing is believing? A beginners' guide to practical pitfalls in image acquisition. JCB vol. 17 no. 1 9–18.Kim, O., MicrobeHunter Microscopy Magazine, An overview of mounting media for microscopy. http://www.microbehunter.com/an-overview-of-mounting-media-for-microscopy/ (accessed 15 July 2014).

CHAPTER THREE

3

Immunohistochemistry and Immunocytochemistry

Simon Renshaw

Abcam plc, Cambridge, UK

IMMUNOCHEMISTRY IS THE DEMONSTRATION of a particular antigen (typically a protein) in a histological tissue section (**immunohistochemistry, (IHC)**) or in a cytological preparation (**immunocytochemistry, (ICC)**) using an antibody that specifically binds to that antigen. The binding of the antibody (commonly termed a 'primary' antibody) is then visualized using an appropriate reporter label, which can be either the precipitated end product of an enzyme/substrate reaction or fluorescent. Reporter labels are often conjugated to a suitable amplification system rather than directly to the primary antibody, in order to increase the signal strength, which is of particular importance when trying to visualize an antigen that is only weakly expressed. Amplification systems can simply consist of a reporter label directly conjugated to a 'secondary' antibody that recognizes and binds to the species and antibody subclass of the primary antibody,

---

This chapter has been modified from *The Immunoassay Handbook*, 4th edition, David G. Wild, Immunohistochemistry and Immunocytochemistry, 357–377, Copyright 2013, with permission of Elsevier.

Immunohistochemistry and Immunocytochemistry: Essential Methods, Second Edition. Edited by Simon Renshaw.

through to more complex amplification systems such as the avidin–biotin complex (ABC) or compact polymer, which offer a greater degree of sensitivity.

This chapter will explore in greater depth the basic concepts mentioned earlier, giving the reader the necessary theoretical and practical knowledge required to perform immunochemical staining procedures with confidence. Particular attention will be paid to quality assurance with regard to the accuracy and reproducibility of results.

Example images are shown in the final section of this chapter.

SPECIMEN FORMATS FOR IMMUNOCHEMISTRY

There are four common specimen formats for immunochemistry, namely paraffin embedded, frozen, free floating and cytological.

Paraffin Embedded

Tissues are harvested from a suitable gross specimen, fixed in order to preserve antigenicity and morphology, and then processed to paraffin wax. Tissue sections are then cut on a microtome, commonly at a thickness of 4 μM (approximately 1 cell thick), with the paraffin wax providing the supporting media in order to make this possible. Sections are then floated out on a water bath at a temperature of 45 °C to help 'iron' out any creases in the tissue generated during microtomy. After being picked up on a glass microscope slide, tissue sections are then dried at 37 °C overnight, in order to aid the section's adherence to the slide. Sections are then de-waxed and have some necessary pre-treatments done before being immunohistochemically stained and microscopically interpreted.

Frozen

Frozen sections are the ideal choice for antigens that do not withstand standard aldehyde fixation regimes and paraffin processing. Antigenicity is therefore preserved to a greater degree in frozen sections, rendering antigens in a more 'native' form. This is due to the absence or relatively low levels of methylene bridge formation (*see* 'The Mechanism of Formaldehyde Fixation') combined with the absence of the harsh chemical and high-temperature environments involved with paraffin processing (*see* 'Processing Tissue Blocks to Paraffin Wax'). Immediately after being excised from the gross specimen, fresh tissue is commonly plunged into a suitable cryo-agent, such as liquid nitrogen-cooled isopentane, in order to freeze the tissue and to reduce the incidence of ice-crystal artefact. The frozen tissue is then placed on a cryostat chuck and embedded at −20 °C in a suitable medium, such as OCT. Cryostat sections can then be cut at around 4–10 μM if destined to be mounted on microscope slides or at around 40 μM if destined to be free floating. Frozen sections that have not originated from aldehyde fixative perfused animals are then briefly fixed in either formaldehyde or a suitable coagulant fixative before being immunohistochemically stained and microscopically interpreted.

Due to the absence of processing to paraffin wax and the frequent absence of optimal aldehyde fixation, the morphology and image resolution of frozen sections is often poorer than in aldehyde-fixed paraffin-embedded tissue sections. However, as long as optimal frozen section preparation techniques are followed, these effects are minimal.

Free Floating

Free-floating sections often originate as stated earlier from frozen sections or can be obtained using unfrozen tissue cut to around 40 μM on a vibratome. These are almost always produced from aldehyde fixative perfused animals and are very common for the visualization of neurological antigens. Such sections are floated out on a suitable buffer solution and immunohistochemical staining performed *in situ*. The thickness allows the sections to survive this process. Sections can then be mounted on microscope slides and microscopically interpreted.

Cytological

Cytological preparations are commonly in the form of conventional smears or spin preparations, or with increasing popularity as 'Thinprep' monolayers on microscope slides. Established laboratory somatic cell lines can also be grown on chamber slides, glass coverslips or 96-well imaging plates. The slides, coverslips or 96-well plates are then briefly fixed in either formaldehyde or a suitable coagulant fixative before being immunocytochemically stained and microscopically interpreted. Cytological preparations generally have the same pros and cons as frozen sections as far as antigenicity is concerned, but it should be remembered that established somatic cell lines are often far removed from native antigen expression when compared to similar cytological preparations from *in vivo* subjects due to a high degree of mutation. An example of this is the comparison of the expression of a particular antigen in HeLa cells with epithelial cells from an endocervical smear.

FIXATION

Tissues destined for immunohistochemistry are collected from deceased or living subjects. These may be of animal or human origin. As soon as the normal *in vivo* homeostatic processes and controls are removed following the subject's death or excision from a living donor, changes occur within the tissue that can have critical effects on antigenicity. Early onset of hypoxia, changes in pH and lysosomal enzymes can have very deleterious effects on antigen preservation. Later stages of degradation involve putrefaction of tissue by bacteria and moulds, leading to gross loss of both antigenicity and cellular morphology. It is therefore imperative that tissues, or more specifically the antigens that are to be demonstrated by immunochemistry, are preserved in an *in vivo* state as much as possible. This is achieved by a process termed 'fixation'. In the interest of optimal antigen preservation, the time between obtaining the tissue and introducing a fixative should be kept to an absolute minimum. It is for this reason that studies involving neurological antigens often employ cardiac gross specimen perfusion to utilize a quick fixation. However, in most cases, fixation is achieved by immersing the excised tissue in a suitable fixative solution that penetrates the tissue and biologically inactivates proteins by creating conformational changes in the tertiary structure.

Types of Fixatives

Fixatives can simply be classified into those that exert their preservative qualities by denaturing proteins (coagulative fixatives) and those that do so by other means, such as covalently cross-linking adjacent proteins (non-coagulative fixatives). A more comprehensive classification of fixatives is provided in Table 3.1.

Aldehyde fixatives (formaldehyde), protein-denaturing agents (methanol, ethanol) and acetone are the most commonly used for immunochemical staining techniques. Far less common are fixatives containing some of the other chemicals mentioned in Table 3.1, for example Bouin's, B5, Zenker's and zinc formalin, each with a specialized application (see p 42).

Formaldehyde

Formaldehyde is a low molecular weight gas (HCHO) with the –CHO being the functional aldehyde. When dissolved in water, these HCHO molecules form methylene hydrate ($HO–CH_2–OH$), which in turn react with each other to form polymers (Fig. 3.1).

Such solutions are termed 'formalin' and consist of 40% (v/v) formaldehyde and 60% (v/v) water. For most immunochemical purposes, working solutions are used at between

TABLE 3.1 Classification of Fixatives

(1) Aldehydes: formaldehyde (paraformaldehyde, formalin); glutaraldehyde; acrolein; glyoxal; formaldehyde mixtures containing mercuric chloride, acetic acid, zinc and periodate lysine
(2) Protein-denaturing agents (precipitants): acetic acid; methanol; ethanol and denatured alcohol
(3) Oxidizing agents: osmium tetroxide; potassium permanganate and potassium dichromate
(4) Other cross-linking agents: carbodiimides
(5) Physical: heat/microwaves
6) Miscellaneous/unknown: non-aldehyde containing fixatives; acetone and picric acid

Source: Ref. [1].

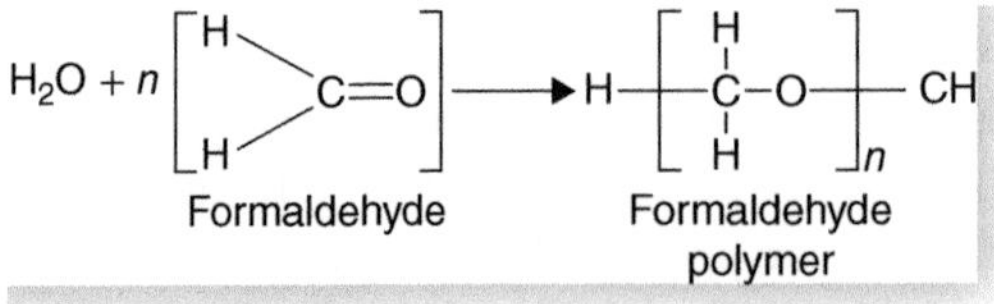

FIGURE 3.1 Formation of Formaldehyde Polymers.
Source: Reproduced with permission of Elsevier

4% and 10% (v/v) formaldehyde content and consist of formaldehyde polymers of 2–8 repeat units. Polymers of up to 100 repeat units are termed 'paraformaldehyde' and are insoluble. However, in order to penetrate tissues and work effectively as a fixative, working formaldehyde solutions need to consist predominantly of monomeric methylene hydrate. This involves the monomerization of the polymerized form. This is commonly achieved by diluting formalin to 10% (v/v) by using a buffer of physiological pH. The transformation is almost instantaneous and is catalysed by hydroxide ions in the slightly alkaline solution. Paraformaldehyde can also be used to generate a working formaldehyde solution of monomeric methylene hydrate. This is achieved by heating the solution to 60 °C with a source of hydroxide ions, commonly by using the salts used to buffer the solution between pH 7.2 and 7.6 (Fig. 3.2).

The Mechanism of Formaldehyde Fixation— The aldehyde groups of formaldehyde form methylene bridge cross-links between lysine resides on the external surfaces of adjacent proteins (Fig. 3.3).

The cytoplasm is therefore transformed into a proteinaceous gel-like network, and the cell is effectively rendered in as much of an *in vivo* state as possible. Soluble proteins become covalently bonded to insoluble proteins. Cellular components such as lipids, carbohydrates

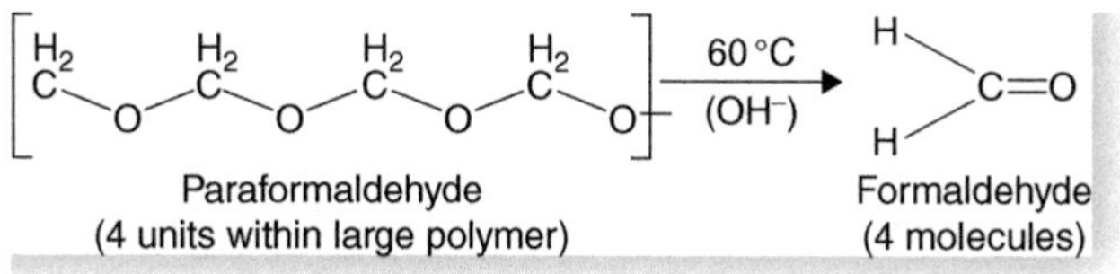

FIGURE 3.2 Monomerization of Paraformaldehyde.
Source: Reproduced with permission of Elsevier

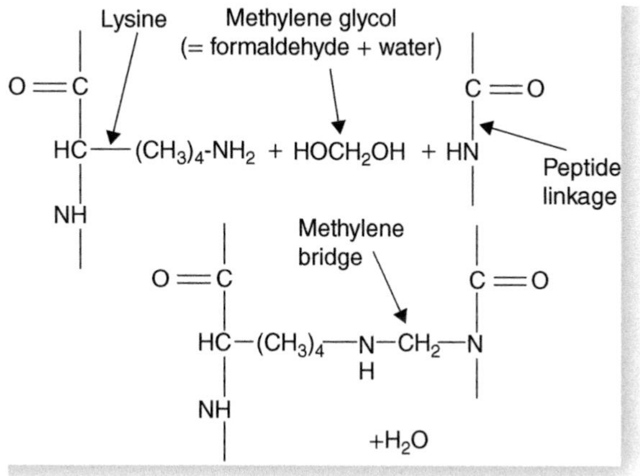

FIGURE 3.3 Formaldehyde Fixation through Cross-Linking of Adjacent Proteins.
Source: Reproduced with permission of Elsevier

and nucleic acids are not directly cross-linked by formaldehyde, but are instead trapped inside the network of cross-links.

Glutaraldehyde

Glutaraldehyde molecules (HCO–$(CH_2)_3$–CHO) are also relatively small, consisting of two aldehyde groups separated by a flexible chain of three methylene bridges (see Fig. 3.4). Glutaraldehyde therefore has double the potential for cross-linking proteins and the flexible CH_2 chain allows this to occur over variable distances.

Glutaraldehyde exists as polymers of various lengths in aqueous solution [2]. Aldehyde groups are present at each end of the polymer, plus one sticking out of each individual polymer unit (see Fig. 3.5). This provides a great ability for binding with lysine molecules, accounting for the rapid fixation of tissue by glutaraldehyde (minutes to hours) compared with that of formaldehyde (see Fig. 3.6).

Free unbound aldehydes are also present that cannot be washed out of the tissue and can provide a source of non-specific antibody binding if not blocked appropriately (see p 64).

As with formaldehyde, glutaraldehyde should ideally be monomeric or at least oligomeric in order to facilitate rapid penetration of tissue. The larger polymers penetrate slowly, and in any case the rapid rate of fixation further reduces the rate of penetration. Similarly, tissues fixed in glutaraldehyde are not penetrated well by paraffin wax, making sectioning difficult. However, resin embedding media penetrate glutaraldehyde-fixed tissues well enough to allow the cutting of ultra-thin sections for electron microscopy [3].

For immunochemical staining, antigen masking is a major drawback with glutaraldehyde fixation due to the excessive and aggressive cross-linking of proteins. Glutaraldehyde

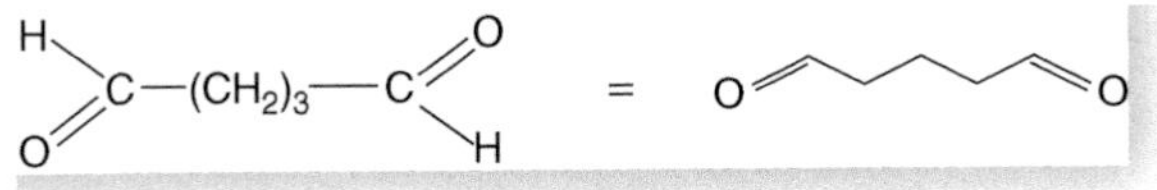

FIGURE 3.4 Chemical Structure of Glutaraldehyde.
Source: Reproduced with permission of Elsevier

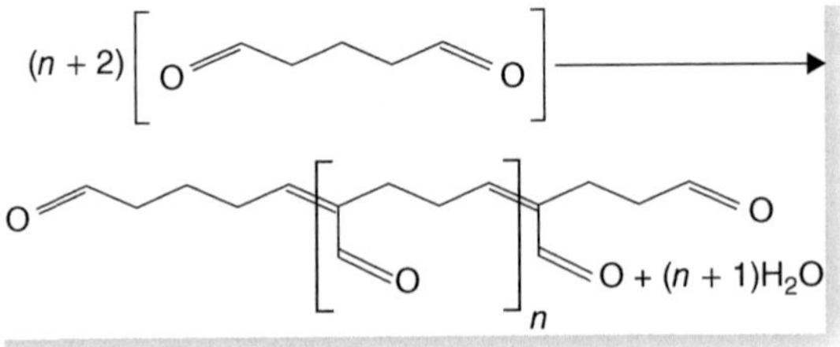

FIGURE 3.5 Glutaraldehyde Polymer Formation.
Source: Reproduced with permission of Elsevier

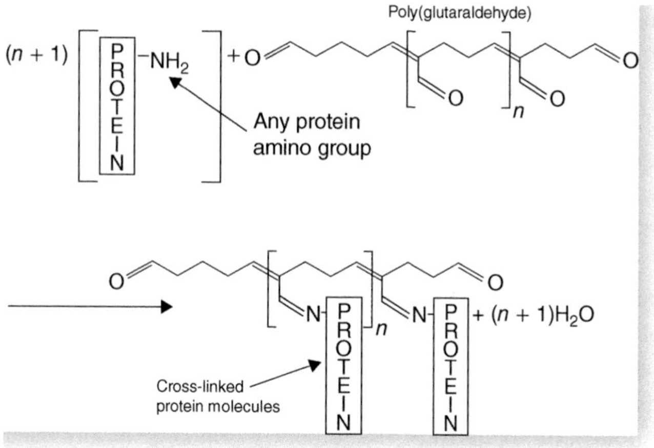

FIGURE 3.6 Mechanics of Glutaraldehyde Methylene Bridge Formation.
Source: Reproduced with permission of Elsevier

is therefore considered largely unsuitable for tissues destined for immunochemical staining and is best used for electron microscopy where a high degree of cytological preservation is of paramount importance. However, small neurotransmitter molecules such as GABA (γ-aminobutyric acid) can be demonstrated immunochemically in tissues by raising antibodies towards such molecules using glutaraldehyde as the protein carrier linker (see p 12). When using such antibodies, inclusion of glutaraldehyde in the formaldehyde fixative is essential, as the epitope comprises the neurotransmitter molecule and the glutaraldehyde molecule. Fixative mixtures of glutaraldehyde and formaldehyde also give a good compromise, allowing fast stabilization of the tissue by the fast-penetrating formaldehyde, combined with rapid fixation from the slow-penetrating glutaraldehyde. Such mixtures are termed Karnovsky's fixatives and typically contain 1–4% (v/v) glutaraldehyde in 2–4% (v/v) formaldehyde [4].

Protein-Denaturing Agents (Precipitants)

Protein-denaturing agents exert their effect by precipitating proteins. Hydrophobic bonds in the protein's interior are disrupted, changing the tertiary protein structure. However, secondary structures are maintained since hydrogen bonds are unaffected and appear to be more stable in alcohols than in water. Since protein-denaturing fixatives do not form cross-links between proteins, they are particularly suited for fixing cytological specimens and frozen sections that are destined for immunochemical staining, as no antigen retrieval step is required. Being friable, such specimens often do not physically survive conventional antigen retrieval techniques. The trade-off is that cytological and morphological details are not as well preserved as with cross-linking fixatives, but this can be beneficial when demonstrating antigens that are sensitive to formaldehyde fixation and tissue processing. If, for whatever reason, formaldehyde is employed to fix frozen or cytological specimens, exposure should be no longer than 5–10 min in order to avoid any high degree of cross-linking.

Ethanol, methanol and industrial methylated spirits (IMS) (70% (v/v) ethanol, 30% (v/v) methanol) are commonly used to fix frozen sections, depending largely on the individual laboratory's personal preference and the antigen being demonstrated. Antigens displaying carbohydrate moieties, such as membrane-bound surface antigens, are commonly fixed with alcohol because alcohol precipitates carbohydrates. However, alcohol fixation tends to hinder (CD) marker staining. Frozen sections are usually fixed for 10–30 min in 70–95% (v/v) alcohols in order to help reduce the morphological distortion of nuclear detail and cytoplasm shrinkage seen with absolute alcohol. Some laboratories use an acetic acid:methanol/ethanol mixture for the same reason [5]. Acetic acid also aids alcoholic penetration of the tissue.

Alcohol (and acetone) penetrates tissue poorly and is generally used only on tissue sections or cytological preparations rather than pieces of tissue (see p 127).

The same protein-denaturing agents used to fix frozen sections can be used to fix cytological preparations, and usually the same principles and considerations apply. There are several commercially available fixative formulations intended especially for cytological specimens containing alcohol (typically 95% (v/v) ethanol) and polyethylene glycol (PEG). The ethanol fixes the cells and the PEG forms a solid hydrophobic layer over them. This layer protects the cells from mechanical damage and stops the ethanol solution from evaporating, giving adequate cellular fixation and preventing the cells from drying out.

Other Fixative

Acetone— For cell surface markers such as the CDs, acetone appears to be the most commonly used fixative. However, extended buffer washes during immunocytochemical staining can create undesirable morphological changes in acetone-fixed tissue such as loss of cell membranes and chromatolysis. Chloroform and certain other desiccants have been added to acetone in order to try and reduce these effects, but with no great degree of success [6]. Fixing frozen sections at room temperature for 10 min in acetone followed by drying at room temperature for 12–48 h has been reported to improve morphology, but increased background staining may occur. Typically, frozen sections are first fixed for 10 min in 100% acetone (after overnight air drying or, when directly removed from the freezer, being allowed to warm to room temperature for 10 min) and are then immediately immunochemically stained (see p 128). Some laboratories include an extra 10 min air-drying step after acetone fixation in order to minimize changes in tissue morphology (see p 128). Extended fixation in acetone can make the tissue brittle.

Periodate/Lysine/Paraformaldehyde— Periodate oxidizes sugars in the tissue to create aldehydes, which are then cross-linked by lysine, while the paraformaldehyde cross-links proteins [7]. Potassium dichromate is a common additive (termed PLDP) to preserve lipids [8]. The three major cellular components are therefore fixed, but the increased degree of cross-linking may accentuate antigen masking.

Bouin's — Caution – picric acid is potentially explosive. Observe appropriate laboratory COSHH (Control of Substances Hazardous to Health) guidelines.

Bouin's fixative is a formaldehyde-based fixative containing saturated picric acid and glacial acetic acid. Fixation times are similar to those for formaldehyde, although fixation in excess of 24 h can cause some tissues to become brittle. It provides a good degree of preservation for glycogen, but may cause tissues to shrink. Picric acid stains the tissue yellow, and after fixation, washes of 50% (v/v) and 70% (v/v) ethanol are used to remove this. Any excess remaining in the tissue sections can be removed with further 70% (v/v) alcohol or 5% (w/v) sodium thiosulphate washes. Bouin's fixative offers excellent morphological preservation and facilitates the removal and chemical alteration of lipids. As this fixative contains formaldehyde, antigen masking may be an issue. Miller [9] claims to have observed reduced immunochemical staining intensity using CD5 (clone 4C7), CD10 (CALLA, clone 56/C6) and cyclin D1 (clone AM29) antibodies on Bouin's fixed tissues.

B5 — B5 is a mercuric chloride/formaldehyde-based fixative. Mercuric chloride is added in order to improve morphological preservation while reducing the distortional changes associated with formaldehyde. Mercuric chloride/formaldehyde-based fixatives do not penetrate tissues well and therefore only small pieces of tissue should be used and fixation periods should be short (2–15 h depending on tissue thickness).

Mercuric chloride/formaldehyde-based fixatives are both cross-linking and coagulative in nature. Their coagulative properties give an exaggerated hardness to tissues, but as these fixatives contain formaldehyde, antigen masking may be an issue. Nuclear and cytoplasmic morphology is well preserved along with good immunolocalization of antigens. Bone marrow and lymph node biopsies are often fixed in B5 for this reason.

Several antigens have been observed to underperform in B5-fixed tissues compared with formaldehyde-fixed tissues. Examples are CD5 (clone 4C7) [10], CD30 and CD23 [9]. Antibodies towards kappa and lambda light chains have been reported to perform better in B5-fixed tissues than in formaldehyde-fixed tissues [9].

Sections fixed in B5 may require de-Zenkerization, which is explained in detail in the following.

Zenker's — Zenker's fixative is also based on mercuric chloride/formaldehyde, but with added potassium dichromate. It essentially performs as for B5, but tissues will require a water wash for 1 h after fixation to remove the potassium dichromate.

Sections fixed in B5 and Zenker's may require de-Zenkerization. This is the term given to the removal of the mercuric chloride deposits from tissue sections. Sections are treated with 70% (v/v) alcohol containing 0.5% (w/v) iodine for 5 min, washed in running water, decolourized in 5% (w/v) sodium thiosulphate for 2 min and finally washed in running water again.

De-Zenkerization is often instructed before immunochemical staining, but it has been demonstrated [9, 11] that this should be performed directly before counterstaining, as de-Zenkerization appears to have a deleterious effect on many antigens. The Lugol's iodine has been shown to exert this effect.

Zinc Formalin— This fixative gives excellent nuclear morphology, and for this reason several authors have claimed it to be preferable to formaldehyde [12–14]. As this fixative contains formaldehyde, antigen masking may be an issue.

Other Beneficial Effects of Fixation

In addition to the preservation of protein, fixation can also improve antibody penetration. The gel-like, porous, proteinaceous network of the cytoplasm facilitates the entry of antibodies. The degree of porosity is largely dependent on the type of fixative used, with cross-linking fixatives giving a reduced degree of porosity compared with coagulative fixatives. Non-coagulative fixatives also denature cell membranes by removing lipids, thus further aiding antibody penetration. Acetone alone is particularly efficient at this and some laboratories add non-ionic detergents (Triton X-100, Tween-20) or saponin to fixative solutions [15] to solvate lipid membranes. This is not recommended, however, for electron microscopy protocols due to the resulting deterioration in subcellular morphology. Also, in certain instances, detergent permeabilization is detrimental to immunochemical staining, as target proteins are lost from the samples, especially when membrane-bound proteins are being demonstrated. Issues regarding antibody penetration are more applicable to cytological specimens, as the cells have not had the added benefit of being bisected by a microtome blade during sectioning.

Fixation also serves to stabilize the specimen and protect it from the physical rigours of processing and immunochemical staining. Aldehyde fixatives, for example, harden tissue and therefore assist with sectioning and survivability during harsh antigen retrieval techniques [16].

Quality Control Considerations Regarding Fixation

Fixation can be considered the very foundation upon which immunochemical staining is based. There is no point in attempting to demonstrate an antigen immunochemically if it has not been adequately preserved. Also, assessing the success of immunochemical staining is based on antigen localization within the tissue, which requires a high degree of tissue morphology preservation. There is therefore a need to obtain a delicate balance between under-fixation and over-fixation, which in practical terms strikes the balance between good tissue morphology and antigen preservation. Factors that significantly affect the degree of fixation are as follows:

- Fixative penetration.
- Fixative concentration.
- Fixation duration and temperature.

The following discussion will highlight the importance of standardized SOPs when performing fixation. It will also highlight the fact that there is no such thing as one ideal fixative solution because some antigens prefer one fixative in preference to another. This is of significant importance when optimizing a new antibody for immunochemistry (*see* p 57).

Penetration

A fixative must obtain rapid penetration into the tissue in order to exert its fixative effects. This is more applicable to formaldehyde, as precipitant fixatives tend to be used only on frozen tissue sections that were not fixed prior to undergoing microtomy, and to cytological preparations. Rapid penetration is facilitated by taking as small a block of tissue as possible from the gross specimen in order to minimize the distance that the fixative needs to penetrate to reach the centre. Current recommendations are for tissue block dimensions to be no greater than $1.0 \times 1.0 \times 0.4$ cm.

Protein-denaturing fixatives penetrate tissue poorly and are therefore often used only on frozen sections or cytological preparations rather than pieces of tissue. Some laboratories use a mixture of 5% (v/v) acetic acid and 95% (v/v) alcohol in order to aid penetration.

Concentration

In general, the higher the fixative concentration, the greater the degree of fixation will be over any given time period.

Working solutions of formaldehyde are typically used at 10% (v/v) on tissues and at around 4–10% (v/v) for cytological preparations.

Ethanol and methanol are commonly used at 10–95% (v/v) on both frozen sections and cytological preparations, in order to help reduce the loss of cytoplasmic detail and nuclear distortion that is often observed with 100% (v/v) alcohol. The addition of acetic acid to the alcohol also helps to prevent this. 100% ethanol and 100% methanol can also be combined in a 1:1 ratio for some antigens.

Acetone is commonly used at 100% (v/v) on both frozen sections and cytological preparations.

Duration and Temperature

Tissue blocks of the dimensions stated earlier should also be fixed for a minimum of 3.5 h and for no longer than 24 h. The degree of methylene bridge formation increases over time. Under-fixation, therefore, results in poor antigen preservation and morphology. Over-fixation results in a high degree of morphology and also a higher degree of antigen masking due to the higher number of methylene bridge cross-links. These methylene bridge cross-links can form a physical barrier to antibody binding, leading to the necessity for antigen retrieval prior to immunochemical staining (*see* 'Antigen Retrieval'). Since all chemical reactions are influenced by temperature, the higher the temperature, the quicker the degree of methylene bridge formation. For tissue blocks, formaldehyde fixation at room temperature for 18–24 h is regarded as a good compromise between adequate morphological and antigenic preservation. For cytological preparations of frozen sections that have originated from unfixed tissue, 10-min formaldehyde fixation at room temperature is suitable. Since the cells or tissue is only around 4 μM thick, penetration is not really an issue. Ten minutes provide an adequate degree of cross-linking without over-fixation, as frozen sections are less likely to survive antigen retrieval than paraffin embedded with cytological preparations even less so.

Protein-denaturing fixatives and acetone are commonly used to incubate frozen sections and cytological preparations for 5–10 min. Frozen sections can be fixed at room temperature. Cytological preparations should be fixed with chilled protein-denaturing agents or acetone at −20 °C in order to help prevent cell dissociation.

Frozen sections that have been fixed with 100% acetone often have a tendency to be brittle and display morphological changes, such as a loss of nuclear membranes. Such changes can be reduced by leaving frozen sections to dry thoroughly overnight before acetone fixation, and then for a further 10 min afterwards, before immunohistochemical staining is performed.

Other Considerations Regarding Aldehyde Fixatives

Fresh solutions should always be prepared with formaldehyde. Solutions of monomeric methylene hydrate slowly revert back over time to the polymerized form; hence, the older the formaldehyde solution, the less effective it is as a fixative. Formaldehyde solutions should also be neutrally buffered at around pH 7.0 because acidic formaldehyde solutions can result in the formation of formalin pigment in tissue sections. This is a brown/black-coloured pigment created from the reaction of the acidic formaldehyde with blood. This can mimic the precipitated end product of the horseradish peroxidase, HRP/DAB reaction, potentially leading to false-positive results. Formaldehyde fixation also hardens tissue, aiding microtomy after processing to paraffin wax.

Other Considerations Regarding Protein-Denaturing Fixatives

No covalent cross-links are formed with protein-denaturing fixatives. No antigen retrieval is subsequently necessary with specimens fixed in protein-denaturing fixatives. Thus, precipitant fixatives are particularly suited to frozen sections and cytological specimens because such specimens tend to be friable and do not withstand heat-mediated antigen retrieval very well. Protein-denaturing fixatives can therefore be employed on frozen tissue sections for antigens that do not tolerate aldehyde fixation or paraffin processing.

For further discussion on fixation, see Bancroft and Gamble [17].

PROCESSING TISSUE BLOCKS TO PARAFFIN WAX

Blocks of tissue that have been optimally fixed in formaldehyde and are destined to be paraffin embedded are required to undergo 'processing'. The paraffin wax is the medium used to support the tissue during microtomy. The basic concept is that tissues are incubated in a series of graded alcohols to replace the water, and then a suitable organic clearing agent such as xylene is used to replace the alcohol, before the tissue is incubated in molten paraffin wax, which in turn replaces the xylene. The methodology is that tissues begin in an aqueous phase (fixed in a solution of formaldehyde). Water is present not only on the outside of the tissue, but throughout. Wax, however, is organic and therefore will not readily mix with the water, the problem being that the wax needs to fully permeate the tissue in order for it to act as a microtomy supporting medium. Alcohol is miscible in both aqueous and organic liquids, so it effectively acts as a bridge between the two phases.

Tissue processing is a poignant consideration for immunohistochemical quality control (QC). Tissue processing takes around 12 h in total. Around 8 h of this is spent with the tissues incubating in alcohol, around 2 h in xylene and then a further minimum of 2 h in molten wax. Alcohol itself is a protein-denaturing fixative, so it will further influence the antigenicity of some antigens. The latter stages of processing occur at around 60 °C, which can degrade some antigens. In summary, some antigens may not survive tissue processing very well, or even not at all. In such cases, a frozen section is far more appropriate for the immunohistochemical demonstration of the antigen because it does not have to undergo the same degree of tissue processing.

Xylene-free processing also exists, but care should be exercised if a laboratory switches from one to the other, since it is feasible that the antigenicity of certain targets may be affected.

For further discussion of processing theory and practice, see Bancroft and Gamble [17].

MICROTOMY

From a QC perspective, both paraffin-embedded and frozen sections are best cut at around 4 μM. With thicker sections, there is an increased possibility for antibodies to get trapped within the tissue, leading to non-specific staining. The exception is frozen sections cut at 40 μM if destined to be free floating.

Tissue sections of 4 μM size, both frozen and paraffin embedded, are usually mounted on glass microscope slides prior to being immunohistochemically stained. Slides should be coated in order to maximize the survival of the tissue during immunohistochemical staining. Poly-L-lysine coated, APES (3-aminopropyltriethoxysilane) coated and positively charged slides are usually available formats. This is of particular importance for paraffin-embedded tissue sections that are to undergo the harsh conditions of heat-mediated antigen retrieval (*see* 'Antigen Retrieval').

For further discussion of microtomy theory and practice, see p 128 and Bancroft and Gamble [17].

TISSUE MICROARRAYS

The **tissue microarray** (**TMA**) is becoming increasingly popular. TMA tissue slides consist of multiple cores of tissue taken from numerous donor blocks and placed into a single recipient block. TMA blocks can therefore consist of tissues from various species, organ types and disease states. They are often custom-built according to the experimental design. In a single immunohistochemical assay, the operator can therefore glean a significant amount of information regarding the binding characteristics of a new antibody or screen numerous tissues for the presence of a certain antigen.

However, since the sample size of each tissue core is very small when compared to that of a single tissue section (typically between 0.6 and 3.0 mm in diameter), great care must be taken to ensure that the cores are representative of the donor block. For example, a block of cancerous tissue may well also contain areas of normal tissue, and random core sampling may produce an area of normal tissue. Due to this, it is not uncommon for two or three cores to be taken from a donor block and placed into the same TMA recipient block. A haematoxylin and eosin (H&E) stained section should therefore be taken of donor tissue blocks and areas of interest microscopically marked out for core sampling. It is also highly beneficial to take H&E sections regularly through the recipient TMA block to ensure that critical lesions are not cut through during microtomy.

For further discussion of TMA theory and practice, see Kumar [18].

SPECIMEN STORAGE

Specimen storage can be split into two categories: pre-immunochemical staining and post-immunochemical staining.

Paraffin-Embedded Tissue Sections

In pre-immunohistochemically stained paraffin-embedded sections, the expression of certain antigens can deteriorate over the period of several months to the point of being immunohistochemically undetectable, depending on the antigen in question (DiVito [19]). This does not seem to be the case for paraffin tissue blocks, which retain antigenicity for many years (Manne [20]). The mechanism is poorly understood, but it is generally believed that oxidation may play a significant role. Tissue sections should not therefore be cut 'en mass' and stored for future use. It is far better to cut them from the paraffin block as and when required. However, dipping paraffin-embedded tissue sections briefly in molten wax after being mounted on a glass slide, and then allowing the wax to set with the slide in a horizontal position, is a proposed way of forming a thin anaerobic barrier over the otherwise exposed tissue. Recently, it has been hypothesized that the presence of water plays a crucial role in antigen degradation. This applies both endogenously from inadequate dehydration during tissue processing and exogenously from damp storage conditions (Xie [21]). This further enforces the importance of tissue processing and the need to minimize storage humidity.

With post-immunohistochemically stained paraffin-embedded sections, storage conditions are usually dependent upon the chromogen and the mounting media used. Sections mounted using adhesive media can be stored easily at room temperature without disrupting the coverslip and tissue section (*see* 'Mounting'). Sections mounted using non-adhesive media have to be stored horizontally, which takes up extra space. They should also be stored at +4 °C in order to prevent mounting media evaporation. The common practice of sealing around the edges of the coverslips with clear nail varnish is a pointless exercise because it is rarely effective. The use of vacuum grease around the edges of coverslips is an effective resolve. Non-adhesive mounting media should therefore be considered as a short-term solution and not for tissue sections destined for archiving. With regard to chromogens, the precipitate resulting from the reaction of HRP with DAB is very resistant to fading in sunlight, whereas that of HRP with AEC fades within a short period of time. Some precipitates also have a tendency to diffuse away from the site of deposition over time, such as that resulting from the reaction of HRP and CN (*see* p 27). It is therefore advisable to store all tissue sections in the dark wherever possible and always check on the chromogen datasheet for information regarding storage characteristics.

Frozen Tissue Sections

Pre-immunocytochemical staining, frozen tissues and their sections should be kept at least at −20 °C, optimally at −80 °C. At no stage should they be allowed to thaw, unless they are tissue sections about to be immediately fixed and immunohistochemically stained.

Post-immunohistochemical staining, frozen sections should follow the same guidelines as for paraffin-embedded tissue sections.

Cytological Specimens

Pre-immunocytochemical staining, it is advisable to store cytological preparations that are adhered to glass microscope slides using one of the many commercially available cytological fixative solutions. These are typically alcohol-based solutions containing PEG. The alcohol acts as a non-coagulative fixative and the PEG forms a waxy layer that covers the cells, protecting them from over-dehydration and from physical damage. Such specimens should be kept at +4 °C.

For fixed cytological preparations grown on chamber slides, in 96-well or glass coverslip plates, filling the wells with a PBS solution containing 0.1% (v/v) sodium azide keeps the cells hydrated and bacteria-free for several months at +4 °C.

Post-immunocytochemical staining, cytological specimens should follow the same guidelines as for paraffin-embedded tissue sections.

Fluorescently Labelled Specimens

Specimens of any nature that have been immunochemically stained using fluorescent reporter labels need to be stored in the dark to protect them from the potential effects of photobleaching. Non-adhesive mounting media containing anti-fade additives should be used (*see* 'Mounting').

DECALCIFICATION

After optimal fixation and before undergoing paraffin processing, some tissues require decalcification. Calcium deposits can build up in some tissues as part of a disease process.

Breast tissue is a common example. Such deposits cause damage to microtome blades and ultimately create scours in tissue sections. Removal of calcium deposits is therefore required. This is achieved by immersing the tissue in a suitable decalcification agent before being processed to paraffin wax.

There are numerous decalcification agents available from proprietary solutions produced by commercial companies to common acids and chelating agents. Nitric acid (5–10%), hydrochloric acid (10%), TCA (trichloroacetic acid; 5%), formic acid (5%) and EDTA (ethylenediaminetetraacetic acid; 10%) are commonly used.

From a QC perspective, it is imperative that the decalcification solution does not destroy the antigen to be immunohistochemically demonstrated. Strong acids, such as nitric and hydrochloric, are very quick to decalcify a tissue (often only a couple of hours), but often give rise to a high degree of antigen loss. It is therefore advised to use a weaker acid, such as formic or TCA. Better still is to use a gentle chelating agent, such as EDTA. However, this process may take several weeks, but it has the advantage of more highly preserved antigenicity.

For further discussion of decalcification theory and practice, see Bancroft and Gamble [17].

ANTIGEN RETRIEVAL

Antigen retrieval (also known as **epitope retrieval** or **unmasking**) serves to reverse the antigen-masking effects of aldehyde fixation (*see* 'The Mechanism of Formaldehyde Fixation'). Methylene bridges between adjacent proteins physically prevent antibodies from accessing the epitope. Some antigens will therefore stain very weakly or even not at all if antigen retrieval is not performed prior to immunohistochemical staining. Some antigens do not seem to require any antigen retrieval at all, but it is generally considered that the demonstration of most antigens does benefit from some degree of antigen retrieval, regardless.

The two common methods of antigen retrieval are heat induced and enzymatic. The exact mechanism of antigen retrieval is poorly understood. It is believed that heat or proteolysis is fundamental to the process by breaking methylene bridges and allowing the antibody access to the epitope. Calcium precipitation is also thought to play a significant role because methylene bridge formation allows bonding of proteins to calcium.

There is no such thing as a universal antigen retrieval solution. Each antigen must be investigated in turn in order to establish the most appropriate antigen retrieval solution (or enzyme), pH, method and duration. As discussed earlier, antigen retrieval success is largely dependent on the degree of methylene bridge formation within tissues (directly linked to the concentration, temperature and duration of aldehyde fixation), the susceptibility of the antigen being demonstrated to antigen masking and the conditions of the antigen retrieval process (temperature, duration, method utilized, and the composition and pH of the antigen retrieval solution). It is therefore imperative from a QC perspective to keep all of the above-mentioned parameters as consistent as possible for any given antigen being demonstrated in order to minimize variability.

It is highly advised to mount tissue sections on charged or coated microscope slides following microtomy. The often harsh conditions of antigen retrieval can dissociate tissues from slides, rendering them useless for subsequent immunohistochemical staining. Poly-L-lysine coated, APES (3-aminopropyltriethoxysilane) coated and positively charged slides are popular choices.

Heat-Induced Antigen Retrieval

Heat-induced antigen (epitope) retrieval (HIER) involves heating tissue sections in a suitable antigen retrieval solution at a certain temperature for a certain duration. Common methods are pressure cooker and microwave based, ranging from domestic units through to purpose-built systems. Purpose-built systems are favourable since they take into account certain inherent problems with the domestic versions. Domestic microwaves, for example, are notorious for creating hot and cold spots within the media they are heating. In context, this means that the degree of antigen retrieval will vary according to where the tissue section is in the vessel containing the antigen retrieval solution. However, scientific microwaves are available that incorporate stirring systems to equilibrate the temperature within the antigen retrieval solution. They also have temperature monitoring systems that hold the antigen retrieval solution just below boiling point so to achieve adequate antigen retrieval while lowering the risk of tissue section dissociation. Scientific pressure cookers tend to have the same feature. In both cases, the operator can usually select the duration and temperature of retrieval. Some scientific microwaves have built-in pressure vessels, hence combining the two systems. Certain automated immunohistochemical staining platforms are available with antigen retrieval systems built in, such as the Leica BOND, which are the ultimate in convenience. Generally, the more automated the antigen retrieval platform is, the more consistent the results will be. Ultimately, the choice of whether to use microwave, pressure cooker or some form of automated HIER seems to be largely dependent on the trends and finances of the individual laboratory.

The success of HIER is largely dependent on the degree of methylene bridge formation within tissues (directly linked to the concentration, temperature and duration of aldehyde fixation), the susceptibility of the antigen being demonstrated to antigen masking and the conditions of the antigen retrieval process (temperature, duration, method utilized, and the composition and pH of the antigen retrieval solution). A difference of 2–3 min of antigen retrieval time can have a significant effect on the intensity of subsequent immunochemical staining. Commonly used antigen retrieval solutions are trisodium citrate buffer (pH 6) and EDTA (pH 9). If all other factors are kept constant, some antigens will immunochemically stain more intensely with one antigen retrieval solution compared to the other, demonstrating the effects of buffer composition and pH.

The advantages of performing HIER far outweigh the disadvantages, if performed correctly. However, there are several factors that the operator needs to be aware of:

- HIER solutions are extremely hot, so great care must be taken from a health and safety perspective, especially when pressurized vessels are involved.
- Since HIER solutions are extremely hot, tissue sections dry almost instantly when removed from them. This 'flash drying' can cause loss of antigenicity and produce artefacts. It is therefore advised to run cold water into the vessel used to hold the antigen retrieval solution until the solution is cool enough for the slides to be moved. Ten minutes in running tap water is good for this, and keeping the process consistent will help to keep results consistent.
- The over-retrieval of antigens can lead to dissociation of tissue section, destruction of the antigen leading to false-negative staining and increase the degree of non-specific/false-positive staining.

Protocol 1 – HIER Protocol: Pressure Cooker Method

Equipment and Reagents

- Domestic stainless steel pressure cooker (preferably a purpose built automated unit, but a domestic one will suffice)
- Hotplate (if using a domestic pressure cooker)
- 1 litre beaker or conical flask
- [a]Trisodium citrate buffer, prepared by mixing: 2.94 g of trisodium citrate, and 1 litre of ultrapure water
- pH meter
- Concentrated hydrochloric acid solution
- Xylene (or other dewaxing reagent)
- 100% IMS or methanol

Method

1. Prepare the trisodium citrate[a] buffer by mixing the trisodium citrate and ultrapure water together in a 1 litre beaker or conical flask. Use a magnetic stirrer to ensure that all reagents are properly dissolved, and pH to 6 with the concentrated hydrochloric acid.
2. Dewax and rehydrate the paraffin sections by placing them in three changes of xylene for 3 min each, followed by three changes of IMS or methanol for 3 min each and then followed by cold running tap water.
3. Add the sodium citrate solution and the dewaxed slides to the pressure cooker[b]. If domestic, place the pressure cooker on the hotplate and turn it on to full power, after securing the lid according to the manufacturer's instruction. If using an automated unit, follow the manufacturers instruction carefully.

4. If domestic, once the cooker has reached full pressure (see manufacturer's instructions), time for 3 min[c]. If automated, set to retrieve antigens for 5 minutes at 110 °C.
5. If domestic, when 3 min has elapsed, turn off the hotplate and place the pressure cooker in an empty sink. Activate the pressure release valve (see the manufacturer's instructions) and run cold water over the cooker.
6. Once depressurized (or once the programmed retrieval cycle has finished, if automated), open the lid and run cold water into the cooker for 10 min. Take care with the hot solution and steam.
7. Continue with an appropriate immunochemical staining protocol.

Notes

[a]Trisodium citrate buffer is used as the antigen retrieval solution in this protocol, but it can be substituted with any suitable buffer at its commonly associated pH.
[b]Use a sufficient volume of antigen retrieval solution to cover the slides by a few centimetres.
[c]The time of 3 min is only suggested as a starting point for the antigen retrieval time. Less than 3 min may leave the antigen under-retrieved, leading to weak staining. More than 3 min may leave the antigen over-retrieved, leading to erroneous staining and also increasing the chances of sections dissociating from the slides. A control experiment is recommended beforehand, where slides of the same tissue section are retrieved for 3, 5, 10 and 20 min before being immunochemically stained to evaluate the optimum antigen retrieval time for the particular antigen being demonstrated. If automated, as well as altering time parameters, the retrieval temperatures of 95 °C, 110 °C and 125 °C should also be evaluated.

Protocol 2 – HIER Protocol: Microwave Method

Slides should be placed in a plastic rack for this procedure.

Equipment and Reagents

- Domestic (850 W) or scientific microwave (preferable)
- 1-L beaker or conical flask
- Microwaveable vessel, either inbuilt or to hold approximately 400–500 mL
- [a]Trisodium citrate buffer, prepared by mixing: 2.94 g of trisodium citrate, and 1 litre of ultrapure water
- pH meter
- Concentrated hydrochloric acid solution
- Xylene (or other dewaxing reagent)
- 100% IMS or methanol

Method

1. Dewax and rehydrate the paraffin sections by placing them in three changes of xylene for 3 min each, followed by three changes of IMS or methanol for 3 min each and then followed by cold running tap water. Keep them in the tap water until step 3.

2. Prepare the trisodium citrate[a] buffer by mixing the trisodium citrate and ultrapure water together in a 1 litre beaker or conical flask. Use a magnetic stirrer to ensure that all reagents are properly dissolved. Adjust to pH 6.0[a] with the concentrated hydrochloric acid solution. Add this solution to the microwaveable vessel[b].
3. Remove the slides from the tap water and place them in the microwaveable vessel. Place the vessel inside the microwave. If a domestic microwave is used, set to full power and wait until the solution comes to boil. Boil for 15 min[d] from this point. If using a scientific microwave, programme it so that the antigen is retrieved for 15 min once the temperature has reached 98 °C[d].
4. When 15 min has elapsed, remove the vessel and run cold tap water into it for 10 min. Take care with the hot solution!
5. Continue with an appropriate immunochemical staining protocol.

Notes

[d]The time of 15 min is only a suggested antigen retrieval time. Less than 15 min may leave the antigen under-retrieved, leading to weak staining. More than 15 min may leave the antigen over-retrieved, leading to erroneous staining and also increasing the chances of sections dissociating from the slides. A control experiment is recommended beforehand, where slides of the same tissue section are retrieved for 5, 10, 15 and 20 min before being immunochemically stained to evaluate optimum antigen retrieval time for the particular antibody being used. If a scientific microwave is being used, as well as altering time parameters, the retrieval temperatures of 95 °C, 110 °C and 125 °C should also be evaluated.

Enzymatic Antigen Retrieval

Enzymatic antigen retrieval has been largely superseded by HIER because HIER is generally easier to work with, has greater batch-to-batch consistency and is applicable to a far greater range of antigens.

Enzymatic antigen retrieval involves incubating tissue sections in a proteolytic enzyme solution at a particular concentration and pH and at the enzyme's V_{max} temperature for a certain duration. This is typically performed in a thermostatically controlled vessel, such as a water bath. The success of enzymatic antigen retrieval is largely dependent on the same parameters as for HIER. Temperature and pH are very critical because this can dramatically affect the proteolytic efficiency of the enzyme and therefore the subsequent degree of antigen retrieval. Common enzymes used for enzymatic antigen retrieval are trypsin, pepsin, pronase and proteinase k.

As with HIER, the over-retrieval of antigens can lead to tissue section dissociation, destruction of the antigen leading to false-negative staining and increase the degree of non-specific/false-positive staining. An additional quality consideration when using enzymes is batch-to-batch variation in proteolytic activity.

Protocol 3 – Enzymatic Antigen Retrieval Protocol

Equipment and Reagents

- Water bath containing two troughs (to contain slide racks)
- [e]Chymotrypsin (type II from bovine pancreas)
- Calcium chloride
- Ultrapure water
- 0.5% (w/v) sodium hydroxide solution
- 0.5% (v/v) hydrochloric acid solution
- Xylene (or other dewaxing reagent)
- 100% IMS or methanol
- Paraffin tissue sections
- pH meter

Method

1. Set the water bath to 37 °C[e]. Add the appropriate amount of ultrapure water to each trough and place the troughs into the water bath[f]. Allow the ultrapure water to warm to 37 °C.
2. Dewax and rehydrate the paraffin sections by placing them in three changes of xylene for 3 min each, followed by three changes of IMS or methanol for 3 min each and then followed by cold running tap water for 3 min.
3. Place the slides in one trough of ultrapure water at 37 °C to warm[g].
4. Remove the other trough and into this dissolve 0.1 g of calcium chloride and 0.1 g of chymotrypsin per 100 mL of distilled water, using a magnetic stirrer to ensure that all reagents are properly dissolved[h].
5. Once dissolved, bring the solution to pH 7.8[e] using the 0.5% (w/v) sodium hydroxide and 0.5% (v/v) hydrochloric acid solutions. Return the trough to the water bath and allow this enzyme solution to reheat to 37 °C[i].
6. Transfer the warmed slides into the enzyme solution for a suggested 20 min[j], then remove the slides and place them under cold running tap water for 3 min[k].
7. Continue with an appropriate immunochemical staining protocol.

Notes

[e]Chymotrypsin is used as the enzyme in this protocol, but it can be substituted with any proteolytic enzyme. Ensure to adjust the reaction conditions accordingly with regard to the V_{max} temperature and optimal pH for the enzyme in question.
[f]Use a sufficient volume of ultrapure water to cover the slides.
[g]Placing cold slides into the enzyme solution will lower the temperature of the solution, thereby reducing enzyme activity and could lead to the antigens being under-retrieved.
[h]Chymotrypsin can be very allergenic. Use a facemask and extraction cabinet for weighing out.

[i]Prepare the chymotrypsin solution as quickly as possible to avoid impairing the activity of the enzyme. Allow this solution to return to 37 °C before introducing the slides.
[j]The time period of 20 min is only suggested as a starting point for incubation time. Less than 20 min may leave the antigen under-retrieved, leading to weak staining. More than 20 min may leave the antigen over-retrieved, leading to erroneous staining and also increasing the chances of sections dissociating from the slides. A control experiment is recommended beforehand, where slides of the same tissue section are incubated in the enzyme solution for 10, 15, 20, 25 and 30 min before being immunochemically stained to evaluate the optimum antigen retrieval time for the particular antigen being demonstrated.
[k]Tap water stops the antigen retrieval process by washing away the enzyme.
For further discussion of antigen retrieval, see Kumar [18].

CONTROLS

It is essential that controls are run in any immunochemical staining assay. Without the appropriate controls, any apparent staining is essentially meaningless. They serve to verify that the staining pattern observed is true, accurate and reliable.

There are two categories of controls used in immunochemistry, namely antigen controls and reagent controls.

Antigen Controls

Positive antigen controls are sections of tissue that are known to contain the antigen being demonstrated. A positive result is therefore expected. Such pieces of tissue have had the demonstration of the particular antigen verified in the past, usually by using an antibody of known specificity (typically a well-established monoclonal) and by someone familiar with the microscopic staining pattern of the antigen in question. Positive controls give assurance that the staining methodology and quality of the detection reagents are sound. It is imperative that positive control sections are treated in exactly the same way as the test tissues to eliminate any variation in results from altered methodology. Positive controls are useful when testing tissue specimens of unknown positivity for a particular antigen, using an antibody of known specificity for that antigen on all of the tissues. They are also valuable for characterizing antibodies of unknown specificity towards a particular antigen, using the same positive control tissue to test all of the antibodies.

Conversely, negative antigen controls are sections of tissue that are known not to contain the antigen being demonstrated. A negative result is therefore expected. Any staining observed in negative control tissue sections therefore has to be from non-specific binding of the primary antibody, from some element of the detection system or from an intrinsic property of the test tissue.

Reagent Controls

Reagent controls serve to ensure that any staining observed in the test tissue is generated from the specific binding of the primary antibody to the antigen that it was raised against, and not from an element of the detection system or from an intrinsic property of the test tissue.

False-positive staining arising from an element of the detection system or from an intrinsic property of the test tissue can easily be determined by replacing the primary antibody with diluent alone. Any subsequent staining must therefore originate from an element other than the primary antibody.

For further discussion on controls, see p 136.

IMMUNOCHEMICAL STAINING TECHNIQUES (OPTIMIZING A NEW ANTIBODY)

Upon receiving a new antibody, a common question in any laboratory is 'Exactly how do I go about immunochemically testing this antibody to obtain optimal results'? In this instance, optimal result means a specific and strong signal from the detection system, in the correct cellular compartment of the correct cell type (or cell line), and from tissue of the correct species and disease state where the antigen in question is expected to be expressed. This would be accompanied by no unwanted background or false-positive staining.

This section deals this process in detail, considering all of the QC critical parameters that have been discussed previously.

Pre-Antibody Purchase/Optimization Research

Questions that need to be answered (at least in part) before purchasing an antibody or designing an optimization experiment are as follows:

- What antigen is the antibody raised against and what sort of immunogen was used to raise the antibody?
- In what tissue has the antibody already been used to successfully detect the antigen? What other tissues are expected to express the antigen in question? Is the antigen expressed in normal tissues or only in certain disease states? Is expression of the antigen gender or species specific, or only present in a certain subset of individuals?
- In what somatic laboratory cell lines has the antibody already been used on to successfully detect the antigen? What other somatic laboratory cell lines are expected to express the antigen in question? Does the cell line need exposing to certain conditions, grown on a certain support matrix (such as collagen), or need treating with a certain chemical in order to express the antigen? Does cell confluency affect expression of the antigen in question (e.g. adhesion proteins exhibiting increased expression when cells touch)?
- What specimen format(s) has the antibody already been successfully used with to demonstrate the antigen (paraffin embedded, frozen, cytological, etc.)?

- In which species has the antibody already been used to successfully detect the antigen in question, and in which species is it expected to cross-react?
- In which cellular compartment is the antigen expected to be expressed? Is it organelle specific?
- What fixative(s) have been used to successfully demonstrate the antibody? What was the concentration and formulation of the fixative solution? How was fixation performed in relation to the specimen, for what duration and at what temperature?
- Has antigen retrieval already been proved necessary to successfully demonstrate the antigen? If so, was it heat mediated or enzymatic? Was it performed in a water bath, microwave, pressure cooker or other vessel? Which buffer solution or enzyme was used, and at what pH and formulation? At what temperature was antigen retrieval performed and for what duration?
- At what concentrations or dilutions has the antibody already been used at to successfully demonstrate the antigen?
- What species is the antibody itself raised in?

If possible, always buy an antibody from a well-established commercial company that is renowned for offering well-characterized antibodies and good after sales technical support. Usually, the more characterization information an antibody's datasheet has, the more faith the end user can have that it is indeed specific for the antigen it was raised against. This applies to all applications present on the datasheet, not just IHC or ICC. Western blot data are particularly useful for this purpose. If there is a single band on the Western blot at the correct molecular weight for the antigen in question, then there is a high chance that if the antibody does prove successful in IHC or ICC, that it will give very specific positive staining with very little or no background.

It goes without saying that, where possible, one should only purchase an antibody where the datasheet states that it has already been successfully characterized in IHC or ICC. However, this is not always possible, and any unknown information can be researched, commonly by using the Internet. For example, if your antibody is monoclonal, a search using the clone number can often reveal scientific papers and images demonstrating successful use of the antibody. Such sources will often provide information regarding tissue used, antigen retrieval and working concentrations/dilutions. Performing general searches for the antigen in question can reveal sources such as bioinformatics databases that harbour useful information regarding tissue expression and subcellular localization. If the amino acid sequence of the immunogen is provided on the datasheet, a 'BLAST' search can be performed to give percentage similarity of the antigen in various species, to help assess potential species cross-reactivity. Contacting the technical services department of the commercial company from whom the antibody was purchased can often prove very useful.

Immunogen design and format are something completely out of the hands of the end user (unless of course you are going to the expense of having a custom antibody produced), but is certainly something to consider. Purified proteins as immunogens tend to generate antibodies with a greater chance of recognizing the native protein rather than a short peptide sequence consisting of up to 20 or so amino acids. This is because purified proteins exhibit tertiary structure, whereas peptides do not. Recombinant proteins tend to be the next best thing to purified proteins, although they may not exhibit the same degree of tertiary structure, because they are biologically manipulated forms of the protein produced by non-native host cells, commonly bacteria. This is not to say that antibodies generated using carefully selected peptides (amino acid sequences that are unique to the target antigen, that are hydrophilic, etc.) are not of any value, since many are indeed excellent. The characterization data displayed on the datasheet for a particular antibody are therefore much more reliable source of information as to whether or not the antibody will be successful in your chosen application. For a more in-depth review of optimal immunogen design, see Chapter 1.

Formulating an Antibody Optimization Experiment

From the information obtained earlier, the end user can now begin to formulate an antibody optimization study. It is worth noting at this point that what works for one laboratory does not necessarily work well for another; hence, a comprehensive study using several expected positive tissues (or cell lines), primary antibody concentrations and antigen retrieval solutions (if appropriate) is highly advised. An example of a comprehensive study is provided in Table 3.2.

This particular study uses various combinations of four concentrations of the primary antibody, two antigen retrieval conditions and two expected positive tissues. Optimal conditions can then be 'tweaked' according to the outcome.

With primary antibodies, it is always advisable to use either a monoclonal antibody or a polyclonal antibody that has been affinity-purified against the immunogen used to raise it. This is to eliminate the presence of contaminating antibodies of different specificity that may be present in the antibody solution.

With secondary antibodies, it is advisable to use one that has been cross-absorbed with at least the species of the tissue section, in order to minimize cross-reaction with any endogenous immunoglobulins within the tissue (*see* 'ABC Immunochemical Staining Protocol, Note B').

It is always advisable to use more than one primary antibody concentration. If any background staining is observed, at just one concentration it is impossible to say whether or not the antibody is simply being used at too high a concentration, or whether it is demonstrating cross-reactivity in some way. If the former is the case, then reducing the concentration will result in a reduction of the staining intensity ratio between the regions of expected positivity when compared to the regions of expected negativity. If the latter is the case, then reducing the concentration will result in the staining intensity ratio of both regions staying relatively the same, but at an equally lower intensity.

The use of more than one tissue that is expected to give a positive result is also encouraged not only to increase the chances of getting a pass but also to take into account genetic variation from donor to donor with regard to antigen expression levels. One example of

TABLE 3.2 Example of Antibody Optimization Experiment

Slide	Primary Antibody Concentration (μg/mL)	Expected Positive Tissue	HIER Antigen Retrieval (20 min)
1	0.1	Tissue A	Sodium citrate, pH 6
2	1	Tissue A	Sodium citrate, pH 6
3	5	Tissue A	Sodium citrate, pH 6
4	10	Tissue A	Sodium citrate, pH 6
5	0.1	Tissue B	Sodium citrate, pH 6
6	1	Tissue B	Sodium citrate, pH 6
7	5	Tissue B	Sodium citrate, pH 6
8	10	Tissue B	Sodium citrate, pH 6
9	0.1	Tissue A	EDTA, pH 9
10	1	Tissue A	EDTA, pH 9
11	5	Tissue A	EDTA, pH 9
12	10	Tissue A	EDTA, pH 9
13	0.1	Tissue B	EDTA, pH 9
14	1	Tissue B	EDTA, pH 9
15	5	Tissue B	EDTA, pH 9
16	10	Tissue B	EDTA, pH 9
17	Reagent control (no primary antibody)	Tissue A	Sodium citrate, pH 6
18	Reagent control (no primary antibody)	Tissue A	EDTA, pH 9
19	Reagent control (no primary antibody)	Tissue B	Sodium citrate, pH 6
20	Reagent control (no primary antibody)	Tissue B	EDTA, pH 9

this is oestrogen receptor expression in breast cancer tissue. Some donors will express this antigen and some will not.

Where possible, always work with absolute concentrations when defining primary antibody working solutions, rather than dilutions. For instance, let us assume that some literature advises the use of the antibody at a 1/1000 dilution. 1 mL of a 1/1000 working solution at a starting concentration of 1 mg/mL would contain 1 μg of immunoglobulin. However, had the antibody been at a starting concentration of 0.5 mg/mL, 1 mL of the same 1/1000 dilution would only contain 0.5 μg of immunoglobulin. However, 1 μg/mL will always be 1 μg/mL, regardless of the starting concentration, which makes reproducibility of results much easier. Obviously, if the antibody is not purified and no starting concentration is stated, then antibody working solution will have to be defined as a dilution.

Choice of Detection System

Detection systems allow the visualization of the primary antibody bound to the antigen being demonstrated. There are several common detection system formats available, offering various degrees of signal amplification. Regardless of format, all detection systems involve the use of a reporter label that can be observed when the specimen is viewed using an appropriate microscope.

Reporter labels can be either enzymatic or fluorescent in nature.

Enzyme Reporter Labels

Reporter labels that are enzymatic in nature produce a stable coloured precipitate at the site of primary antibody binding when exposed to a suitable chromogen. The principle of the reaction is as follows:

$$\text{Enzyme} + \text{substrate(chromogen)} \rightarrow \text{enzyme} + \text{product(precipitate)}$$

Among several labels, the two most popular enzyme labels for immunochemistry are HRP and alkaline phosphatase (AP). For an in-depth discussion regarding HRP, AP and their chromogens, see Chapter 2.

When selecting an enzymatic reporter label, one should consider the nature of the tissue being used in the immunostaining experiment, with regard to endogenous enzymes. Endogenous peroxidase is found in red blood cells and can react with the HRP chromogen to produce false-positive staining. However, such activity is usually blocked by applying a solution of hydrogen peroxide to the tissue (*see* 'ABC Immunochemical Staining Protocol, Note H'). Similarly, endogenous AP activity is commonly found in the large intestine. Such activity can be blocked by incubating the tissue in a levamisole solution (*see* 'ABC Immunochemical Staining Protocol, Note J'). However, a more practical solution in this case would be to simply use HRP instead of AP.

Fluorescent Reporter Labels

Fluorescent reporter labels (fluorochromes) are chemical molecules that have the ability to absorb light of a certain wavelength and then re-emit light at a longer wavelength. For an in-depth discussion regarding fluorescent reporter labels, see Chapter 2.

The choice of fluorochrome to demonstrate the binding of the primary antibody largely depends on the filter sets on the end user's microscope and their choice of fluorescent counterstains (*see* 'Counterstains'). Fluorochromes and the associated fluorescent counterstains in a single immunofluorescence (IF) experiment should be selected so that their absorption and emission spectra do not overlap and that each can be excited and observed separately. There are several spectral analyser programmes available, typically on the websites of commercial reagent companies that allow the end user to check the absorption and emission characteristics of the fluorescent labels they intend to use and to ensure that there is no overlap. Such programmes also tend to allow the end user to input the filter sets on their microscope and to further check overall compatibility. In experiments where several fluorescent reporter labels or counterstains are being employed, fluorophores with the highest quantum yields and extinction coefficients should be reserved for the least abundant of antigens, to avoid weaker signals being 'quenched' by stronger signals (see Chapter 2).

Enzymatic or Fluorescent?

In general, enzymatic reporter labels and tinctorial counterstains are used on tissue sections, either formalin fixed, paraffin embedded or frozen. Fluorescent reporter labels and counterstains are usually used on frozen tissue sections and cytological preparations. However, this trend is not set in tablets of stone, and the detection system should be tailored to suit the specifics of the particular immunostaining experiment.

For most, the main benefit of fluorescence over enzymatic is that all of the fluorescent channels can easily be viewed separately and then merged to form a pseudocoloured image. It is therefore easy to see signal co-localization between a certain fluorescent counterstain and that of the detection system, therefore adding credibility to the specificity of the primary antibody. An additional benefit to the fact that each channel is viewed separately is that very weak staining from the primary antibody can be observed in isolation without any interference from other signals. However, tissue sections tend to display a degree of autofluorescence due to various tissue components being naturally fluorescent, such as collagen. Formaldehyde fixation also increases the degree of autofluorescence. This autofluorescence, if strong enough, can mask the signal from fluorescent reporter labels, making the interpretation of fluorescence results difficult. It is for this reason that enzymatic detection is often more appropriate for tissue sections. Cytological preparations and frozen sections are commonly not exposed to formaldehyde for long enough to exacerbate autofluorescence, and many cytological preparations often do not possess such naturally fluorescent components. Most autofluorescence occurs in the green range of the spectrum, and such problems can often be overcome by swapping reporter labels that emit light in the green range (e.g. Alexa Fluor® 488) with one that emits in the red range (e.g. Alexa Fluor® 647).

Signal Amplification

Whatever be the choice of the label, it is always advisable to use some sort of signal amplification rather than a directly conjugated primary antibody, therefore greatly enhancing the visualization of weakly expressed antigens.

Signal amplification typically involves the use of a reporter label-conjugated secondary antibody. The secondary antibody is raised against (and will therefore bind to) the species immunoglobulin subclass of the primary antibody. Secondary antibodies can either be directly conjugated to the label in some way or be conjugated to biotin, allowing the

TABLE 3.3 Detection Systems

Detection System	First Step	Second Step	Third Step
Label-conjugated primary antibody	Label-conjugated primary antibody	N/A	N/A
Label-conjugated secondary antibody	Primary antibody	Label-conjugated secondary antibody	N/A
Dextran polymer	Primary antibody	Dextran-label polymer-conjugated secondary antibody	N/A
Compact polymer	Primary antibody	Label-polymer secondary antibody	N/A
Avidin–biotin complex (ABC)	Primary antibody	Biotin-conjugated secondary antibody	Label-conjugated avidin–biotin complex

subsequent addition of an ABC bearing the conjugated label. For the beginner, seeing immunostaining diagrams can often cause confusion because they are nearly always depicted as 1 primary antibody binding to the antigen, with 1 secondary antibody binding to the primary and with 1 molecule of label conjugated to the secondary antibody. However, in reality, multiple primary antibodies may bind to the antigen (depending on the nature of both the antibody and the antigen) and then multiple secondary antibodies will bind to each of the primary antibodies. Each of the secondary antibodies may then have as many as eight molecules of reporter labels conjugated to them. Thus, at each stage of the immunostaining protocol, provision is being made for increasing the number of labels that are ultimately going to be localized to the binding sites of the primary antibodies. Common detection system regimes (with varying associated degrees of signal amplification) are provided in Table 3.3.

Out of these, always strive to use either dextran polymer, compact polymer or ABC because they give the greatest degree of signal amplification. They are all available from numerous reagent manufacturers, often in a ready-to-use kit format. Polymer systems are now becoming extremely popular because they give a comparable degree of signal amplification when compared to ABC systems, but with the added advantage of being only of two steps, hence quicker and easier to use. Compact polymer is more favourable than dextran polymer, since with compact polymer the label molecules are covalently bonded to each other in close proximity to the secondary antibody, rather than being conjugated to a large dextran–polymer backbone, which is in turn conjugated to the secondary antibody. Due to their increased size, it has been reported that dextran–polymer systems can suffer from steric hindrance problems when trying to bind to the primary antibody and when demonstrating certain intra-cellular antigens. Polymer systems also have the added advantage of being biotin-free, hence tending to give cleaner background staining in certain tissues than ABC. Endogenous biotin in the tissues such as liver can bind to the avidin–biotin complex in ABC systems and give false-positive staining.

For further discussion on signal amplification, see Kumar [18].

Immunostaining Techniques

The following immunostaining protocols are fairly generic and have been shown to deliver reliable and consistent results. They differ slightly as to whether the end user is performing immunostaining on tissue sections or cytological preparations. The differences are outlined in the 'Notes' Section proceeding each protocol. The end user should therefore take the relevant pieces of information to formulate a protocol tailored to their individual needs. If the end user is using a commercial immunostaining kit, the manufacturer's instructions should be observed.

In the interest of quality assurance, ensure that all reagents are freshly made up on the day of the experiment. Allow all frozen or refrigerated reagents to reach room temperature before using them. Ensure that reagents are accurately made using correctly calibrated measurement equipment and that they are adequately mixed. When using reagents purchased from commercial companies, observe the manufacturer's instructions and take note of expiry dates.

Please note that if fluorescent reporter labels are being used, protect all of the incubation steps involving the reporter label from light as far as practically possible to minimize the potential for photobleaching. Thankfully, for second-generation fluorophores such as the Alexa Fluor® range, on-the-bench photobleaching is not really a problem, but it is still a good practice to take minimal light exposure.

Protocol 4 – Avidin–Biotin Complex (ABC) Immunochemical Staining Protocol (Figure 3.7)
Method

Day 1

1. Perform any necessary specimen pre-treatments (dewaxing of tissue sections, antigen retrieval, etc.).
2. Rinse the slides or cells for 3 × 5 min in buffer-containing surfactant[a].
3. Block in buffer-containing surfactant, incorporating 10% (v/v) normal serum, 1% (w/v) BSA and 0.3 M glycine for 2 h at room temperature[b].
4. Rinse the slides or cells for 3 × 5 min in buffer-containing surfactant[c].
5. Apply optimally diluted primary antibody in buffer-containing surfactant, incorporating 1% (w/v) BSA[d].
6. Incubate overnight at 4 °C[e].

Day 2

7. Rinse the slides or cells for 3 × 5 min in buffer-containing surfactant.
8. Apply optimally diluted secondary biotinylated antibody in buffer-containing surfactant, incorporating 1% (w/v) BSA, for 1 h at room temperature (at this point make up the ABC following the manufacturer's instructions)[f,g].
9. Rinse the slides or cells for 3 × 5 min in buffer-containing surfactant.
10. If using an enzymatic label on tissue sections only: perform endogenous enzyme quenching[h], followed by rinsing for 3 × 5 min in buffer containing surfactant.
11. Apply ABC for 30 min at room temperature (made up according to the manufacturer's instructions).
12. Rinse the slides or cells for 3 × 5 min in buffer-containing surfactant.
13. If using an enzymatic label only: develop with the chromogen for 10 min at room temperature (or until the desired degree of staining is achieved as determined under the microscope)[i,j].
14. If using an enzymatic label only: rinse in running tap water for 5 min.
15. Counterstain[k].
16. Dehydrate, clear (if required) and mount[l].
17. There should now be a visual label localized at the site of antibody binding. This corresponds to the location of the target[m].

Notes

[a]It is recommended to use TBS containing 0.025% (v/v) Triton X-100 surfactant for tissue sections and PBS containing 0.1% (v/v) Tween surfactant for cytological preparations. TBS gives a cleaner background than PBS because of its higher salt concentration. However, it can cause cytological preparations to lyse because of this, so it is the best practice to use it on tissue sections only. PBS should also be avoided when using an AP enzymatic label, as phosphate buffers can quench AP activity. Phosphate is an AP substrate, so it will compete with the chromogen.

Surfactants predominately serve to improve antibody penetration into the specimen and work by removing lipid from cell membranes. For some cell membrane proteins, it may be necessary not to use surfactant in the buffer because this may result in the protein being washed away. The use of surfactant also helps to reduce surface tension, allowing the reagents to spread out and cover the specimen with ease. It is also believed to dissolve Fc receptors in frozen sections, thereby helping to reduce specific but undesired background staining.

If a higher degree of permeabilization is required in cytological preparations, then an additional 10-min incubation in 0.1% (v/v) Triton X-100 can be incorporated into the protocol between steps 1 and 2, with an additional buffer wash step before step 2. Triton X-100 is a harsher surfactant than Tween, but this brief exposure to Triton X-100 can be beneficial for the demonstration of many antigens in cytological preparations. About 4 mM sodium deoxycholate is another ionic detergent that permeates the cells by solvating lipid membranes and can be incorporated into the protocol in the same procedure as Triton X-100.

[b]Normal serum should be from the species in which the secondary antibody was raised. For instance, if the primary antibody is raised in rabbit and the secondary antibody is a goat anti-rabbit antibody, then normal goat serum should be used. Normal serum helps to prevent the secondary antibody from cross-reacting with endogenous immunoglobulins in the tissue. For example, when detecting an antigen in human tissue and when the secondary antibody is raised in a goat,

primarily block using normal goat serum. This will bind to any human immunoglobulins that show cross-reactivity with goat immunoglobulins. On addition of the primary antibody, for example a rabbit antibody of IgG isotype, it will bind to its intended target. On addition of the secondary antibody, for example a biotinylated goat anti-rabbit IgG antibody, it will only bind to the primary rabbit IgG antibody because if the secondary antibody did have an affinity for any of the endogenous immunoglobulins in the human tissue, then it can no longer bind to them as they have already been bound to immunoglobulins in the goat serum. Immunoglobulins from the same species will not interact with each other and therefore the secondary antibody can only bind to the primary antibody.

In addition, antibodies are one of the most hydrophobic of the major serum proteins (which is why they tend to form aggregates when stored over time). The higher the protein's degree of hydrophobicity, the higher its likelihood of linking to another particularly hydrophobic protein; thus hydrophobic interactions between tissue proteins and antibodies can occur. This effect is further exacerbated by aldehyde fixation, which is where the glycine helps, because it reacts with and effectively caps off any free aldehyde groups. Non-specific binding of the secondary antibody is therefore prevented by antibodies in the normal serum binding to and effectively blocking the hydrophobic binding sites in the tissue (since antibodies from the same species will not cross-react). BSA also serves the same purpose.

The use of normal serum before the application of the primary antibody also helps to eliminate leucocyte Fc receptor binding of both the primary and secondary antibodies.

[c]A higher degree of blocking may be achieved by simply removing the excess serum (by aspirating or tipping it off) rather than washing it off completely. However, this is optional.

[d]The primary antibody will target the epitope it is raised against. For proteins, this is an amino acid sequence as unique as practically possible to the protein in question. Make sure that the primary antibody is raised in a different species to the tissue being stained. If, for example, mouse tissue is being used and the primary antibody is raised in a mouse, the application of an anti-mouse IgG secondary antibody would bind to all of the endogenous IgG in the mouse tissue and non-specific background would occur.

[e]Overnight incubation allows antibodies of lower titre or affinity to be used by simply allowing more time for the antibodies to bind. Also, whatever be the antibody's titre or affinity for its target, once the tissue has reached saturation point, no more binding can take place. Overnight incubation assures that this occurs. The lower incubation temperature is believed to help reduce background staining by increasing reaction times and therefore favouring antibody-antigen interaction, rather than antibody-background. This can be further enhanced by gentle agitation from an orbital shaker. If time is of the essence, apply the primary antibody for only 1 h at room temperature. An overnight incubation is not usually essential, but has the advantages outlined earlier.

[f]The secondary antibody recognizes the immunoglobulin species and subtype of the primary antibody. In this example, a biotinylated goat anti-rabbit IgG antibody is being used to bind to the primary rabbit IgG. The secondary antibody is biotinylated, meaning that it has been conjugated with biotin (see note [g]).

[g]The ABC consists of biotin–enzyme or biotin-fluorescent label conjugates bound to avidin. When applied, the ABC will bind to the secondary biotinylated antibody. After the addition of the secondary antibody, make up the ABC according to the manufacturer's instructions and leave to stand for a minimum of 30 min at room temperature. This is the length of time that the complex takes to form.

Avidin is a protein found in chicken egg white and has similar properties to streptavidin, a protein found in *Streptomyces avidinii*. Both avidin and streptavidin have a high affinity for biotin, a co-factor in enzymes involved in carboxylation reactions. Both avidin and streptavidin can be used in an immunochemical staining application, but streptavidin tends to be favoured as it shows greater sensitivity. Streptavidin also produces less non-specific background staining, as, unlike avidin, it is not glycosylated, so it shows no interaction with lectins or other carbohydrate-binding proteins.

The ABC is formed through avidin having four binding sites for biotin. The ABC binds to the biotinylated secondary antibody, which is bound to the primary antibody, which is in turn bound to the target on the tissue section.

[h]H_2O_2 suppresses endogenous peroxidase activity and therefore reduces background staining because endogenous peroxidases as well as the peroxidase label, HRP, would react with the chromogen. Using a low concentration of 1.6% (v/v) H_2O_2, in buffer-containing surfactant, for 30 min at room temperature (or 5 min for a frozen section) adequately blocks endogenous peroxidase activity without having a detrimental effect on tissue epitopes. If using AP, then omit this step and step 12. See note [j] for further details on blocking endogenous AP activity.

It is essential that fresh H_2O_2 is used, as H_2O_2 readily breaks down into water and oxygen at room temperature, rendering it ineffective at blocking endogenous peroxidase activity. H_2O_2 is best stored frozen and thawed shortly before use. AP is therefore an ideal label to use when staining tissue high in endogenous peroxidases, such as the spleen.

[i]Develop the coloured product of the enzyme label with the appropriate chromogen. The choice of chromogen depends on the enzyme label being used, the preferred coloured end product and whether aqueous or organic mounting media are being used.

[j]Ensure that any chromogen is made up correctly according to the manufacturer's instructions.

If using DAB, do not forget that it is a suspected carcinogen. Wear the appropriate protective clothing. Deactivate it with chloros in a sealed container overnight (it produces noxious fumes when chloros is added) and dispose of it according to laboratory COSHH guidelines. Appropriate COSHH guidelines should be observed regarding the storage, use, handling and disposal of any laboratory reagents. If using AP, add 0.24 mg/mL levamisole (Sigma L9756) to the chromogen solution. Levamisole suppresses endogenous phosphatase activity and therefore reduces background staining, although not in the placenta or small intestine. The AP label will not be affected by levamisole.

[k]If using an enzymatic label, apply the appropriate nuclear counterstain, commonly haematoxylin. The desired level of staining intensity is largely down to personal preference, but there should be a balance between obtaining a good degree of nuclear morphology while not inadvertently masking any nuclear staining. Try applying the haematoxylin to the specimen for 1 min before differentiation and/or blueing, and adjust accordingly from the results obtained.

If using a fluorescent label, numerous fluorescent counterstains can be used, and these should be chosen according to the guidelines discussed in the section entitled 'Counterstains'. For example, if a 488 label is being used to visualize the primary antibody, then two counterstains could possibly be DAPI to visualize the cell nuclei and Alexa Fluor 594 conjugated to wheat germ agglutinin to visualize the cell membrane. Always use counterstains according to the manufacturer's instructions. It may be necessary to perform optimization experiments to find the most appropriate concentrations.

[l]If using an enzymatic label and chromogen combination where the end precipitate is alcohol soluble (e.g. HRP-AEC or AP-Fast Red TR), employ a suitable aqueous mounting medium. Do not dehydrate and clear before mounting, or the end precipitate will be dissolved!

If using an enzymatic label and chromogen combination where the end precipitate is not alcohol soluble (e.g. HRP-DAB or AP-New Fuchsin), dehydrate and clear by processing them through three changes of 100% (v/v) alcohol for 3 min each and then followed by three changes of xylene (or, in the interest of health and safety, a commercial xylene-alternative clearing agent) for 3 min each. Mount the sections in a suitable organic mounting medium, such as DPX.

With a fluorescent label, do not dehydrate and clear. Use an aqueous mounting medium, as discussed in the section entitled 'Mounting'. PBS containing 10% (v/v) glycerol is a good general mounting medium for fluorescent reporter labels.

[m]Enzymatic immunochemical staining should be viewed using a conventional light microscope.

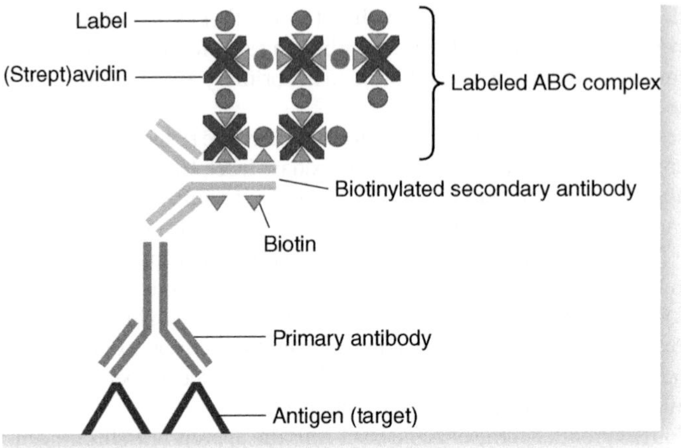

FIGURE 3.7 Avidin–Biotin Complex (ABC) Immunochemical Staining Protocol.
Source: Reproduced with permission of Elsevier

Fluorescent immunochemical staining must be viewed using a fluorescence microscope set up according to the excitation and emission characteristics of the label(s) being used.

Protocol 5 – Label-Conjugated Secondary Antibody Immunochemical Staining Protocol (Figure 3.8)

Please refer to the relevant notes in Protocol 4 for each of the steps and to the special considerations for fluorescent labels. Ensure that the secondary antibody is directly conjugated to the label!

Method

Day 1

1. Perform any necessary specimen pre-treatments (dewaxing of tissue sections, antigen retrieval, etc.).
2. Rinse the slides or cells for 3 × 5 min in buffer-containing surfactant[a].
3. Block in buffer-containing surfactant, incorporating 10% (v/v) normal serum, 1% (w/v) BSA and 0.3 M glycine for 2 h at room temperature[b].
4. Rinse the slides or cells for 3 × 5 min in buffer-containing surfactant[c].
5. Apply optimally diluted primary antibody in buffer-containing surfactant, incorporating 1% (w/v) BSA[d].
6. Incubate overnight at 4 °C[e].

Day 2

7. Rinse the slides or cells for 3 × 5 min in buffer-containing surfactant.
8. If using an enzymatic label on tissue sections only: perform endogenous enzyme quenching[h], followed by rinsing for 3 × 5 min in buffer containing surfactant.

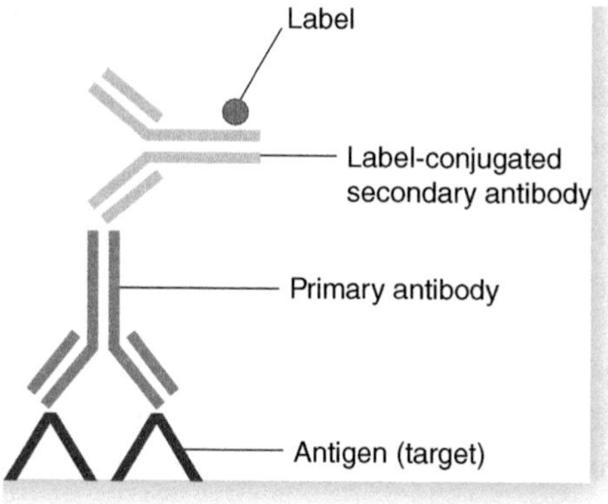

FIGURE 3.8 Label-Conjugated Secondary Antibody Immunochemical Staining Protocol. *Source*: Reproduced with permission of Elsevier

9. Apply optimally diluted label-conjugated secondary antibody in buffer-containing surfactant, incorporating 1% (w/v) BSA, for 1 h at room temperature[f].
10. Rinse the slides or cells for 3 × 5 min in buffer-containing surfactant.
11. If using an enzymatic label only: develop with the chromogen for 10 min at room temperature (or until the desired degree of staining is achieved as determined under the microscope)[j].
12. If using an enzymatic label only: rinse in running tap water for 5 min.
13. Counterstain[k].
14. Dehydrate, clear (if required) and mount[l].
15. There should now be a visual label localized at the site of antibody binding. This corresponds to the location of the target[m].

Protocol 6 – Compact Polymer Immunochemical Staining Protocol (Figure 3.9)
Please refer to the relevant notes in Protocol 4 for each of the steps and to the special considerations for fluorescent labels. Also observe the additional note for this specific protocol.

Method

Day 1

1. Perform any necessary specimen pre-treatments (dewaxing of tissue sections, antigen retrieval, etc.).
2. Rinse the slides or cells for 3 × 5 min in buffer-containing surfactant[a].
3. Block in buffer-containing surfactant, incorporating 10% (v/v) normal serum, 1% (w/v) BSA and 0.3 M glycine for 2 h at room temperature[b].
4. Rinse the slides or cells for 3 × 5 minutes in buffer-containing surfactant[c].
5. Apply optimally diluted primary antibody in buffer-containing surfactant, incorporating 1% (w/v) BSA[d].
6. Incubate overnight at 4 °C[e].

Day 2

7. Rinse the slides or cells for 3 × 5 min in buffer-containing surfactant.
8. If using an enzymatic label on tissue sections only: perform endogenous enzyme quenching[h], followed by rinsing for 3 × 5 min in buffer containing surfactant.
9. Apply optimally diluted compact polymer label-conjugated secondary antibody in buffer-containing surfactant, incorporating 1% (w/v) BSA, for 1 h at room temperature[f,n].
10. Rinse the slides or cells for 3 × 5 min in buffer-containing surfactant.
11. If using an enzymatic label only: develop with the chromogen for 10 min at room temperature (or until the desired degree of staining is achieved as determined under the microscope)[j].
12. If using an enzymatic label only: rinse in running tap water for 5 min.
13. Counterstain[k].
14. Dehydrate, clear (if required) and mount[l].
15. There should now be a visual label localized at the site of antibody binding. This corresponds to the location of the target[m].

Additional Note

[n]Multiple molecules of polymerized immunolabels are directly conjugated to the secondary antibodies, hence avoiding potential steric hindrance problems that can be associated with polymer systems utilizing dextran backbones. Potential non-specific binding is also claimed to be reduced due to the absence of the polymer backbone and better diffusion of the reagents to the target sites.

Immunochemical staining using multiple primary antibodies and reporter labels.

Immunostaining using multiple primary antibodies and reporter labels on the same specimen can provide valuable information regarding the expression of two or more antigens in relation to each other. The above immunochemical staining protocols have been written assuming that only one primary antibody and one reporter label are being used. However, they can easily be adapted to accommodate multiple staining techniques.

For further discussion on multiple staining techniques, see Chapter 4.

Immunochemical staining using a primary antibody raised in the same species as the specimen being stained.

If the primary antibody is raised in the same species as the specimen being stained and a secondary antibody is being used, then high background staining will occur because the secondary antibody will bind to not only the primary but also any endogenous immunoglobulins in the specimen. An example of this is using a mouse primary antibody on mouse tissue. Of course, the simplest solution is to use a primary antibody raised in a species with sufficient phylogenetic difference to that of the specimen, such as a rabbit primary on human tissue (a mouse primary antibody on rat tissue, for example, may

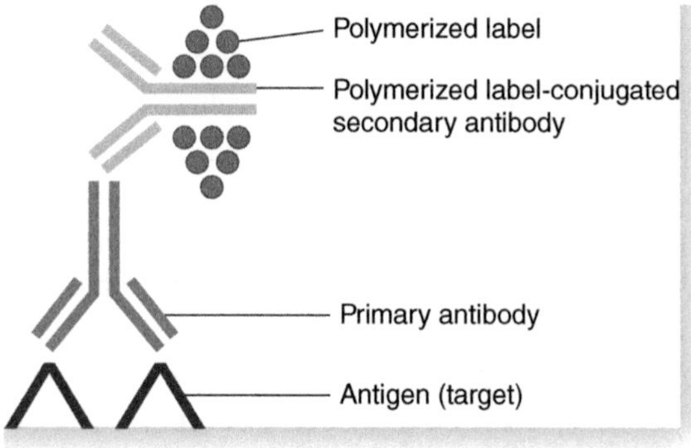

FIGURE 3.9 Compact Polymer Immunochemical Staining Protocol.
Source: Reproduced with permission of Elsevier

still generate background staining if it cross-reacts with rat immunoglobulin). However, if this is not possible, then one of the simplest ways to avoid this is to directly biotinylate the primary antibody (using a commercially available biotinylation kit) and use an ABC system as a second-step detection reagent. If the antigen in question is abundant, then the primary antibody may also directly be reporter label conjugated. These methods avoid the use of a secondary antibody, while still providing a sufficient degree of signal amplification. Also, there are many commercial 'mouse-on-mouse' kits available, which either involve a biotinylation step or incorporate some other proprietary blocking step.

COUNTERSTAINS

Counterstains serve to add colour contrast to a tissue section or cytological preparation, by specifically staining certain organelles or cellular compartments, thus further defining the localization of the primary antibody and the subsequent protein of interest. They can be either tinctorial or fluorescent in nature to match the nature of the detection system used to visualize the primary antibody.

For enzyme/chromogen detection systems, it is common for a single nuclear counterstain to be used. With immunofluorescence, it is common for both a nuclear and cell membrane counterstains to be used. However, regardless of how many counterstains are used in an immunostaining experiment, they all need to be of a different colours (or have sufficiently difference absorption and emission characteristics if fluorescent) to both themselves and the detection system used so that the signals can be sufficiently distinguished from each other.

Counterstains for Enzyme/Chromogen Immunostaining

Haematoxylin is arguably the most common nuclear counterstain used when employing an enzyme/chromogen detection system. There are various formulations available, classified by the type of mordant used and whether they are progressive or regressive. All of them ultimately give cell nuclei a pleasing blue colouration of varying hue and intensity depending on the type of haematoxylin used.

Haematoxylin alone (or more accurately its oxidation product, haematin) is anionic and therefore does not have much affinity for DNA. Mordants are iron salts, namely those of iron, aluminium, tungsten and lead. Mordants combine with haematin resulting in a positively charged dye–mordant complex, thus allowing it to bind to anionic chromatin. Alum (aluminium mordanted) haematoxylins can be used progressively or regressively. With progressive haematoxylins (such as Mayer's, Carazzi's and Gill's), tissue or cells are incubated in haematoxylin until the desired degree of nuclear staining is achieved before being blued. In comparison, in the case of regressive haematoxylins (such as Harris's), tissues or cells are incubated until a degree of overstaining is achieved, before having some of the excess haematoxylin removed by immersion in an acidic solution, such as 1% acid alcohol. This process is known as differentiation. Progressive haematoxylins are therefore more convenient to use than regressive, due to the absence of a differentiation step and the resulting compatibility with alcohol-soluble enzyme/substrate end products, such as those produced by HRP and AEC. The use of a non-alcohol-containing haematoxylin, such as Mayer's, would be more preferable in this situation.

Whether progressive or regressive, once the desired level of nuclear staining is achieved, haematoxylins are 'blued'. At acid pH, haematoxylins stain the nuclei red. However, once exposed to an alkaline environment, haematoxylin turns into a pleasing blue colour. Running tap water is commonly used for this purpose because it has sufficient alkalinity, especially in 'hard' water areas. In areas of 'soft' water, a suitable alkaline solution can be used to blue haematoxylin, such as 0.05% (v/v) ammonia.

Staining intensity depends on several factors, namely the concentration of the haematoxylin solution, the duration of staining, the concentration of the differentiation solution and the duration of differentiation. The desired level of staining intensity is largely down to personal preference, but there should be a balance between obtaining a good degree of nuclear morphology and not inadvertently masking any nuclear antibody staining.

Other commonly used tinctorial nuclear counterstains are light green, fast red, toluidine blue and methylene blue, staining nuclei either in green, red or blue, respectively.

Counterstains for Fluorescent Immunostaining

There are many commercially available fluorescent counterstains available, specifically engineered to stain many different organelles and cellular structures. They may be chemicals themselves (such as DAPI or Hoechst to stain nuclei) or a fluorescent molecule conjugated to a suitable lectin that has affinity for a particular cellular component (such as wheat germ agglutinin conjugated to Alexa Fluor 594, for staining the cell membrane). Fluorescent counterstains should be selected according to the filter sets on the operator's microscope and the absorption and emission spectra of the fluorescent reporter label(s) being utilized in the staining experiment (see Chapter 2). Table 3.4 shows examples of

TABLE 3.4 Common Fluorescent Nuclear and Cell Membrane Counterstains

Fluorescent Counterstain	Maximum Absorption (nm)	Maximum Emission (nm)	Colour	Target Organelle
Phalloidin, Alexa Fluor® 350	346	442	Purple/blue	Cytoskeleton
Wheat germ agglutinin, Alexa Fluor® 350	346	442	Purple/blue	Cell membrane
DAPI	359	461	Blue	Nucleus
Hoechst	351	461	Blue	Nucleus
Phalloidin, Alexa Fluor® 488	495	519	Green	Cytoskeleton
Wheat germ agglutinin, Alexa Fluor® 488	495	519	Green	Cell membrane
DCS1	503	526	Green	Nucleus
Phalloidin, Alexa Fluor® 555	555	565	Green/yellow	Cytoskeleton
Wheat germ agglutinin, Alexa Fluor® 555	555	565	Green/yellow	Cell membrane
Phalloidin, Alexa Fluor® 594	590	617	Orange/red	Cytoskeleton
Wheat germ agglutinin, Alexa Fluor® 594	590	617	Orange/red	Cell membrane
Phalloidin, Alexa Fluor® 647	650	665	Red	Cytoskeleton
Wheat germ agglutinin, Alexa Fluor® 647	650	665	Red	Cell membrane
DRAQ5™	646	681, 697	Far red	Nucleus

some of the common fluorescent nuclear, cytoskeletal and cell membrane counterstains. This list is by no means exhaustive!

DAPI binds selectively to double-stranded DNA with no or very little cytoplasmic staining. Green or red fluorescent labels can easily be used with DAPI, as there is no significant overlap in the excitation and emission spectra. DAPI can be used on both fixed and unfixed cytological preparations and tissue sections.

Hoechst dyes bind specifically to A/T-rich regions of double-stranded DNA, with no or very little cytoplasmic staining. Their characteristics are very similar to those of DAPI.

With DRAQ5™, blue or green fluorescent labels can easily be used, since there is no significant overlap in the excitation and emission spectra. DRAQ5™ can be used on both fixed and unfixed cytological preparations and tissue sections.

Wheat germ agglutinin selectively binds to sialic acid residues on cell membranes, giving a general cell membrane counterstain. It can be purchased conjugated to a variety of Alexa Fluor® labels.

Phalloidin is a bicyclic peptide isolated from the toxins of the *Amanita phalloides* (more commonly known as 'Death Cap') mushroom. It selectively binds to the cytoskeletal element F-actin (filamentous). Note that methanol fixation does not preserve F-actin well.

Although not specifically mentioned on Table 3.4, antibodies towards tubulin, when directly conjugated to a suitable fluorescent label, such as one from the Alex Fluor® range, make a most beautiful and elegant cytoskeletal counterstain. However, care should be taken to ensure that the anti-tubulin antibody does not cross-react with any unwanted proteins or any subsequently used secondary antibodies.

For all fluorescent counterstains, follow the manufacturer's instructions for the staining procedure.

For further discussion on counterstains, see Kumar [18] and Bancroft and Gamble [17].

MOUNTING

Mounting is the preparation of an immunochemically stained specimen for the analysis of results and subsequent archiving. It serves to protect the specimen and immunochemical staining from physical damage and enhances the quality of the analysis of results by creating a sharper image down the microscope.

For tissues and cytology specimens immunochemically demonstrated using enzymatic reporter labels, this involves covering the specimen with a thin piece of glass known as a coverslip. The coverslip is held in place with a suitable adhesive known as the **mounting medium**. Mounting media are commercially available in numerous formulations, all of them having different properties that need to be optimally matched to the specimen type and reporter label used.

Mounting media should have the following characteristics (adapted from Woods and Ellis [5]):

1. Be colourless and transparent.
2. Completely permeate and fill the tissue interstices.
3. Have no deleterious effects on the tissue.
4. Resist bacterial contamination.

5. Protect the tissue and staining from mechanical and chemical damage (oxidation and pH changes).
6. Be miscible with the dehydrant or clearing agent.
7. Set without crystallizing, cracking or shrinking (or otherwise deforming the material being mounted) and not react with, leach, or induce fading in stains and reaction products (including those from enzyme histochemical, hybridization and immunohistochemical procedures).
8. Once set, the mountant should remain stable (in terms of the features listed earlier). This is particularly important when long-term specimen storage is required.

Simply speaking, mounting media can either be organic or aqueous in nature. Where possible, it is advisable to use an organic mounting medium, such as dibutylphthalate xylene (DPX), since they have a refractive index (RI) close to that of fixed tissue (approximately 1.53), giving a high degree of contrast and sharpness down the microscope. As light passes from one medium to another, it bends and changes speed. This can be practically observed as the apparent bending of a stick when it is placed into water. Light travels fastest in a vacuum and slower in all other media. The RI of any given medium is a ratio comparison of the speed of light in a vacuum (always 1) to that of the medium (always higher than 1). Therefore, the closer the RI of the mounting media to that of fixed tissue the better because the light will be less distorted as it passes through.

However, in order for tissue to be mounted in organic media, it first has to be dehydrated and cleared. The basic concept is that tissues are incubated in three changes of 100% alcohol for 5 min each to replace the water, then for three changes of 5 min each in a suitable organic clearing agent such as xylene to replace the alcohol, before being mounted in an organic mounting medium. The term 'clearing' comes from the fact that xylene has a similar RI to that of fixed tissue, rendering tissue transparent when immersed in it. The methodology is that immunohistochemical staining occurs in an aqueous phase. Water is present throughout the tissue section and if it is still present when mounted, the organic mounting medium will not readily mix with the water. Water droplets will clearly be seen on the tissue section when microscopically examined and macroscopically give the overall specimen a 'milky' appearance. Alcohol is miscible in both aqueous and organic liquids, so effectively it acts as a bridge between the two phases.

Great care should be taken with regard to the choice of mounting medium and the chromogen used. For example, the red-coloured precipitate of the reaction of AEC chromogen with HRP is soluble in alcohol, so organic mounting media would not be a good choice in this case because the dehydration stage would remove it from the tissue! Always check the datasheet for the chromogen used in order to see the nature of compatible mounting media. In cases such as AEC, the only suitable solution is to use an aqueous mounting medium.

Aqueous mounting media have the advantage that, as immunohistochemical staining is also performed in an aqueous phase, they can be used immediately afterwards. This makes them quick and convenient, but since they have a RI further away from that of fixed tissue than organic mounting media, they tend to give a less sharp image down the microscope.

Organic and aqueous mounting media can further be classified as adhesive or non-adhesive. Generally, organic mounting media tend to be adhesive and aqueous non-adhesive. Adhesive mounting media set hard, firmly adhering the coverslip to the specimen. This offers a high degree of immunohistochemical staining preservation and assists the archiving of tissue slides. Non-adhesive mounting media stay as a liquid and often need to be sealed around the edge of the coverslip with vacuum grease if the tissues are not going to be microscopically analysed straight away, to prevent the specimen from drying out.

For tissues and cytology specimens immunochemically demonstrated using fluorescent reporter labels, non-adhesive aqueous mounting media are commonly used. PBS containing 10% (v/v) glycerol is a good general mounting medium for fluorescent reporter labels. Commercially available mounting media usually contain anti-fade chemicals that are free-radical scavengers to help to slow down the effects of photobleaching. Some may also contain nuclear counterstains, such as DAPI. However, care should be taken not to use mounting media that contain glycerol when using phycobiliprotein reporter labels, since it has a quenching effect on the fluorescence.

With immunofluorescence, it is a good idea to microscopically observe the mounting medium of choice on a blank coverslip to ensure that it does not produce any degree of autofluorescence.

For a general discussion of dehydrating, clearing and mounting theory and practice, see Bancroft and Gamble [17].

TROUBLESHOOTING

Troubleshooting immunochemical staining procedures is not an easy task. With so many variables, it is often difficult to know exactly where to begin. The best approach is to start with the simple approach and progress to the more technically demanding issues later. The importance of using appropriate controls cannot be overstressed.

For further information on controls and troubleshooting, see Chapter 5.

EXAMPLES OF IMMUNOSTAINING PHOTOMICROGRAPHS

In this section, photomicrographs showing examples of the capability of immunocytochemistry/immunofluorescence (fourteen images) and immunohistochemistry (eleven images) are provided.

Immunocytochemistry/Immunofluorescence

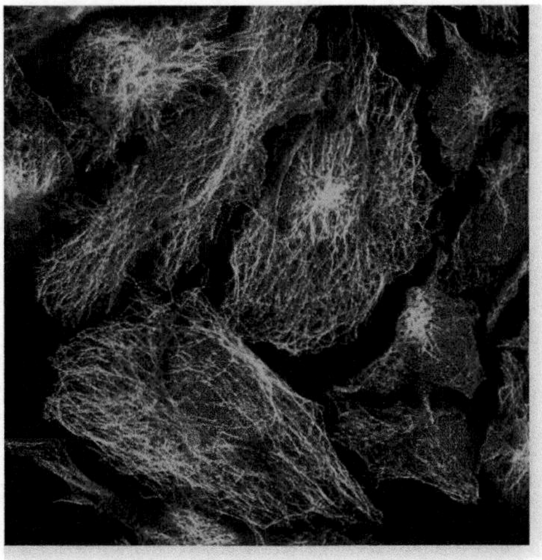

ab195887 staining alpha tubulin (microtubule marker) in HeLa cells. The cells were fixed with 4% formaldehyde (10 min), permeabilized in 0.1% PBS-Triton X-100 for 5 min and then blocked in 1% BSA/10% normal goat serum/0.3 M glycine in 0.1% PBS-Tween for 1 h. The cells were then incubated with ab195887 at 1/167 dilution (shown in green) overnight at +4°C. Nuclear DNA was labelled in blue with DAPI. This product gave a positive signal in 100% methanol (10 min) fixed HeLa cells under the same testing conditions

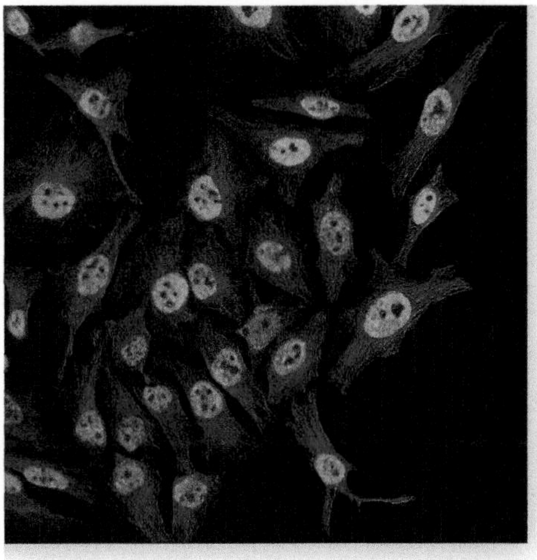

ab207014 staining STAT6 in HeLa cells. The cells were fixed with 4% formaldehyde (10 min), permeabilized with 0.1% Triton X-100 for 5 min and then blocked with 1% BSA/10% normal goat serum/0.3 M glycine in 0.1% PBS-Tween for 1 h. The cells were then incubated overnight at +4°C with ab207014 (Alexa Fluor® 488) at a 1/100 dilution (shown in green) and ab195884, rat monoclonal to alpha tubulin (Alexa Fluor® 647), at a 1/250 dilution (shown in red). Nuclear DNA was labelled with DAPI (shown in blue). This product also gave a positive signal under the same testing conditions in HeLa cells fixed with 100% methanol (5 min)

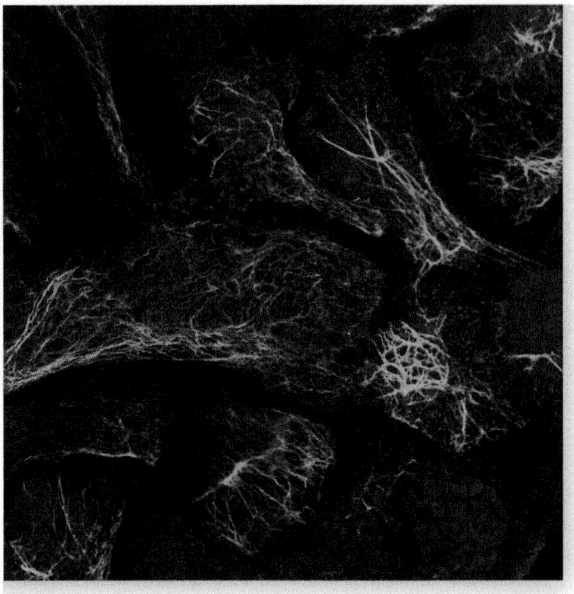

ab207351 staining cytokeratin 5 (intermediate filament marker) in A431 cells. The cells were fixed with 100% methanol (5 min), permeabilized with 0.1% Triton X-100 for 5 min and then blocked with 1% BSA/10% normal goat serum/0.3 M glycine in 0.1% PBS-Tween for 1 h. The cells were then incubated overnight at +4°C with ab207351 (Alexa Fluor® 488) at a 1/100 dilution (shown in green) and ab195889, mouse monoclonal to alpha tubulin (Alexa Fluor® 594), at a 1/250 dilution (shown in red). Nuclear DNA was labelled with DAPI (shown in blue). This product also gave a positive signal under the same testing conditions in A431 cells fixed with 4% formaldehyde (10 min)

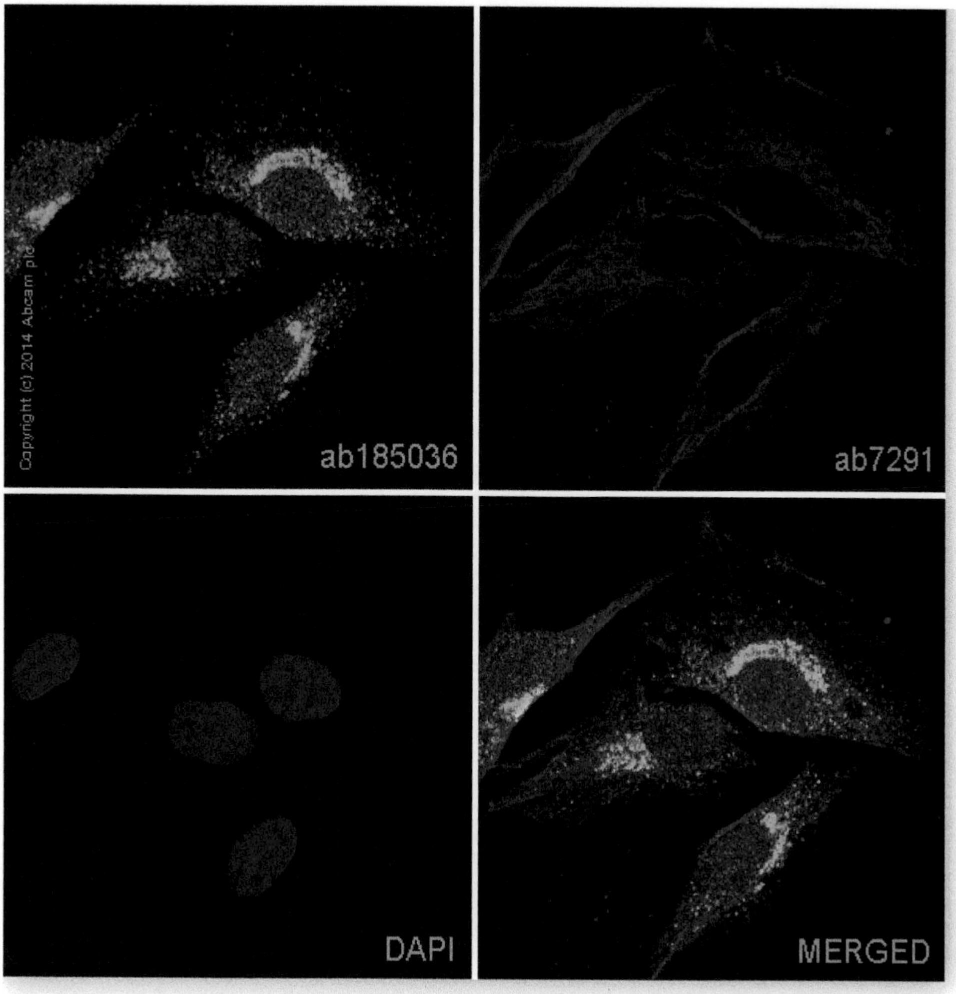

ab185036 staining MAP1LC3A (autophagosome marker) in HeLa cells. The cells were fixed with 4% formaldehyde (10 min), permeabilized in 0.1% Triton X-100 for 5 min and then blocked in 1% BSA/10% normal goat serum/0.3 M glycine in 0.1%PBS-Tween for 1 h. The cells were then incubated with ab185036 (Alexa Fluor® 488) at a working dilution of 1 in 100 (shown in green) and ab7291 (mouse monoclonal [DM1A] to alpha tubulin) at 1 μg/ml overnight at +4°C, followed by a further incubation at room temperature for 1 h with an Alexa Fluor® 594 goat anti-mouse secondary (ab150120) at 2 μg/mL (shown in red). Nuclear DNA was labelled in blue with DAPI

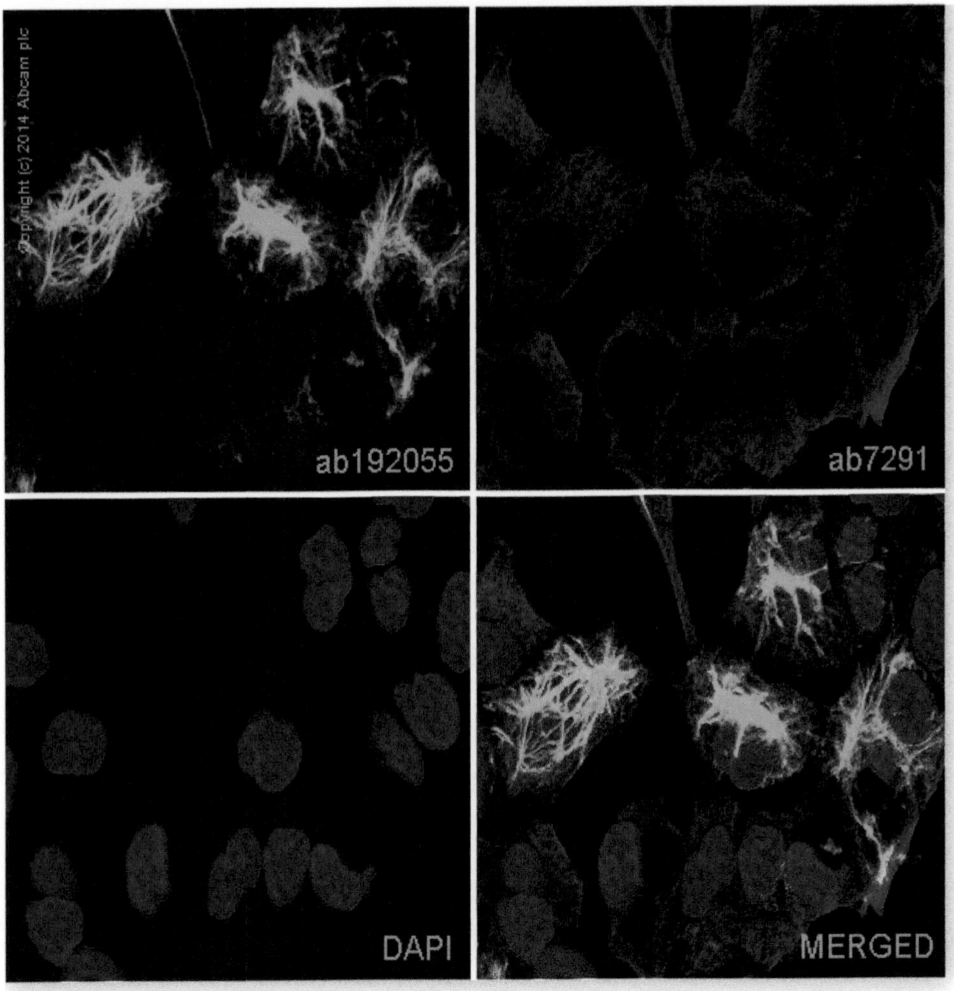

ab192055 staining cytokeratin 14 (intermediate filament marker) in A431 cells. The cells were fixed with 100% methanol (5 min), permeabilized in 0.1% Triton X-100 for 5 min and then blocked in 1% BSA/10% normal goat serum/0.3 M glycine in 0.1% PBS-Tween for 1 h. The cells were then incubated with ab192055 (Alexa Fluor® 488) at a working dilution of 1 in 100 (shown in green) and ab7291 (mouse monoclonal [DM1A] to alpha tubulin) at 1 µg/ml overnight at +4°C, followed by a further incubation at room temperature for 1 h with an Alexa Fluor® 594 goat anti-mouse secondary (ab150120) at 2 µg/mL (shown in red). Nuclear DNA was labelled in blue with DAPI. This product also gave a positive signal in 4% formaldehyde (10 min) fixed A431 cells under the same testing conditions

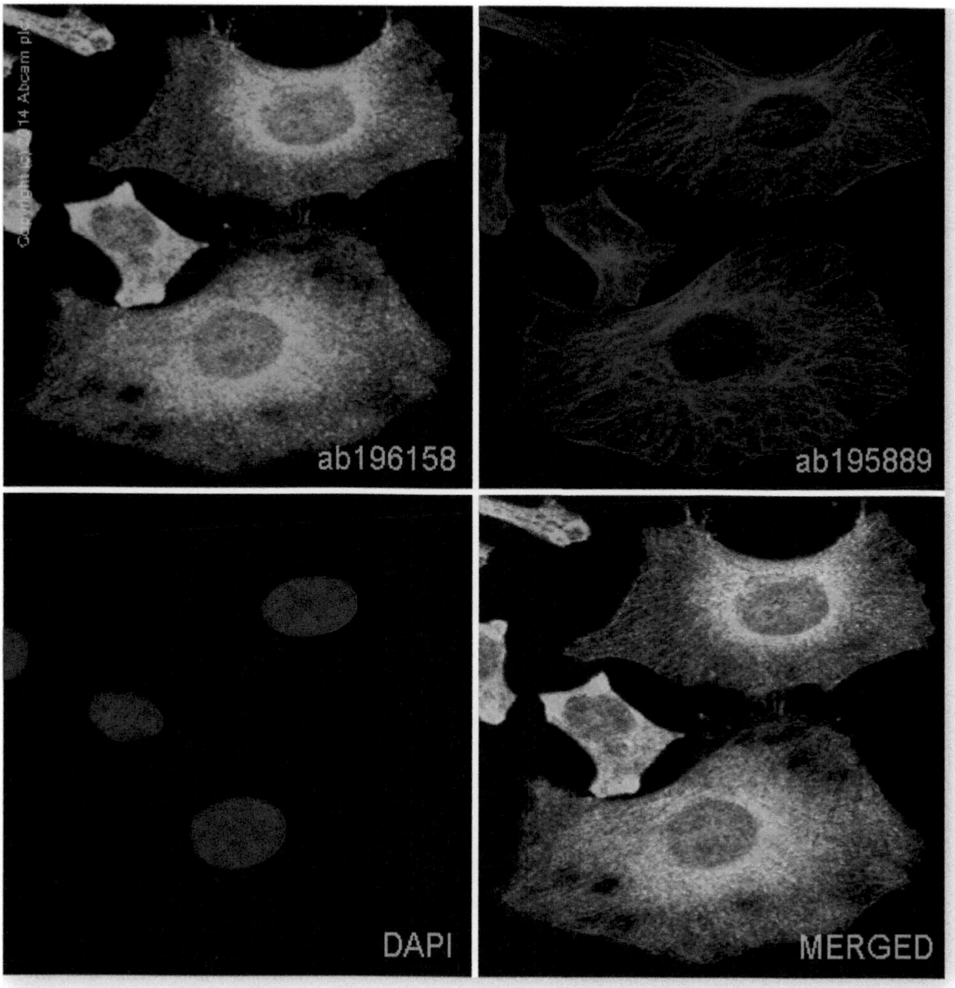

ab196158 staining calreticulin (endoplasmic reticulum marker) in HeLa cells. The cells were fixed with 4% formaldehyde (10 min), permeabilized in 0.1% Triton X-100 for 5 min and then blocked in 1% BSA/10% normal goat serum/0.3 M glycine in 0.1% PBS-Tween for 1 h. The cells were then incubated with ab196158 (Alexa Fluor® 488) at 1/50 dilution (shown in green) and ab195889, mouse monoclonal [DM1A] to alpha tubulin (Alexa Fluor® 594, shown in red) at 2 µg/mL overnight at +4°C. Nuclear DNA was labelled in blue with DAPI. This product also gave a positive signal in 100% methanol (5 min) fixed HeLa cells under the same testing conditions

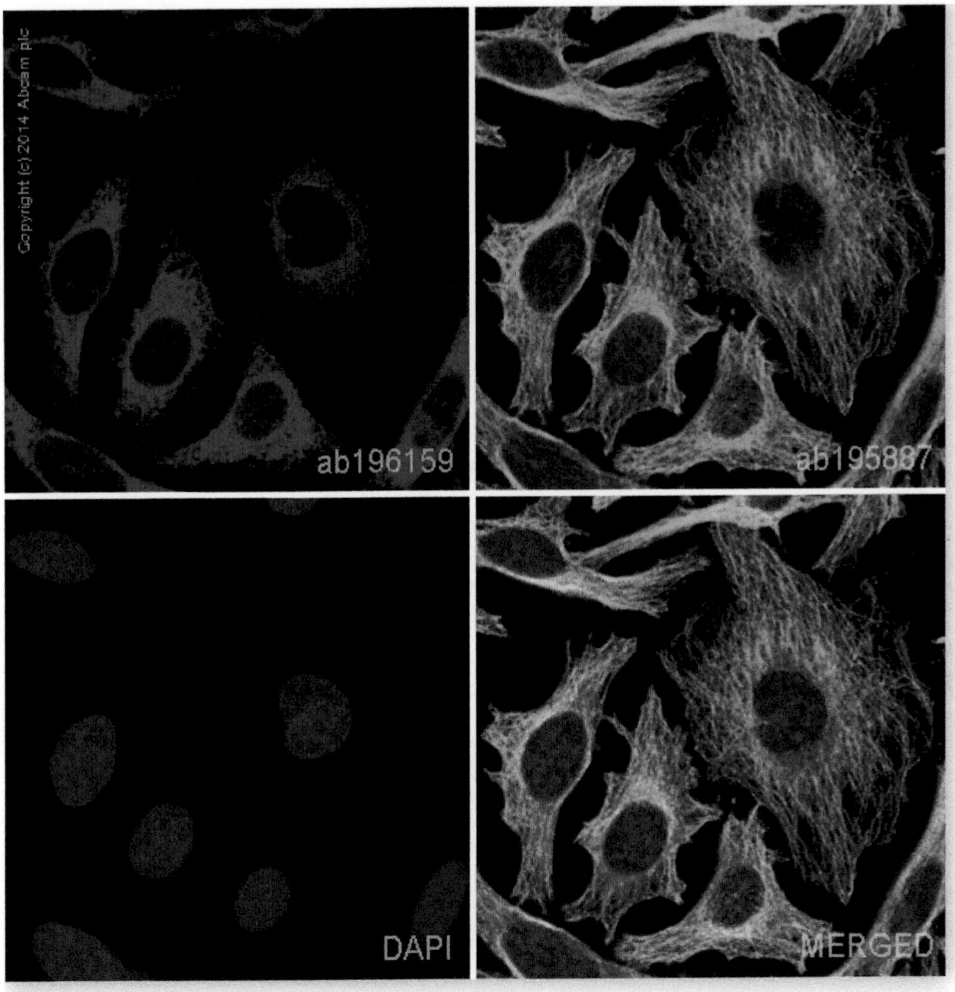

ab196159 staining calreticulin (endoplasmic reticulum marker) in HeLa cells. The cells were fixed with 100% methanol (5 min), permeabilized in 0.1% Triton X-100 for 5 min and then blocked in 1% BSA/10% normal goat serum/0.3M glycine in 0.1% PBS-Tween for 1 h. The cells were then incubated with ab196159 (Alexa Fluor® 647) at 1/100 dilution (shown in red) and ab195887, mouse monoclonal [DM1A] to alpha tubulin (Alexa Fluor® 488, shown in green) at 2 µg/mL overnight at +4°C. Nuclear DNA was labelled in blue with DAPI

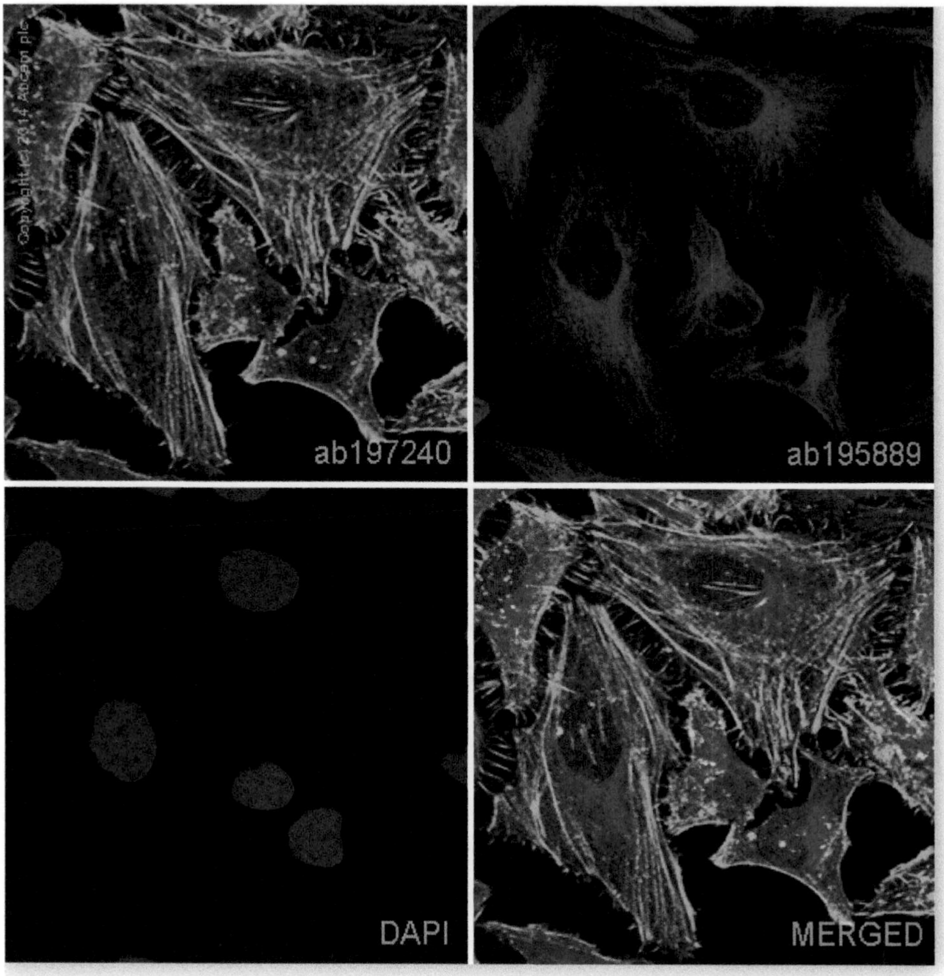

ab197240 staining alpha smooth muscle actin (microfilament marker) in HeLa cells. The cells were fixed with 4% formaldehyde (10 min), permeabilized with 0.1% Triton X-100 for 5 min and then blocked with 1% BSA/10% normal goat serum/0.3 M glycine in 0.1% PBS-Tween for 1 h. The cells were then incubated overnight at +4°C with ab197240 (Alexa Fluor® 488) at a 1/100 dilution (shown in green) and ab195889, mouse monoclonal to alpha tubulin (Alexa Fluor® 594), at a 1/250 dilution (shown in red). Nuclear DNA was labelled with DAPI (shown in blue). This product also gave a positive signal under the same testing conditions in HeLa cells fixed with 100% methanol (5 min)

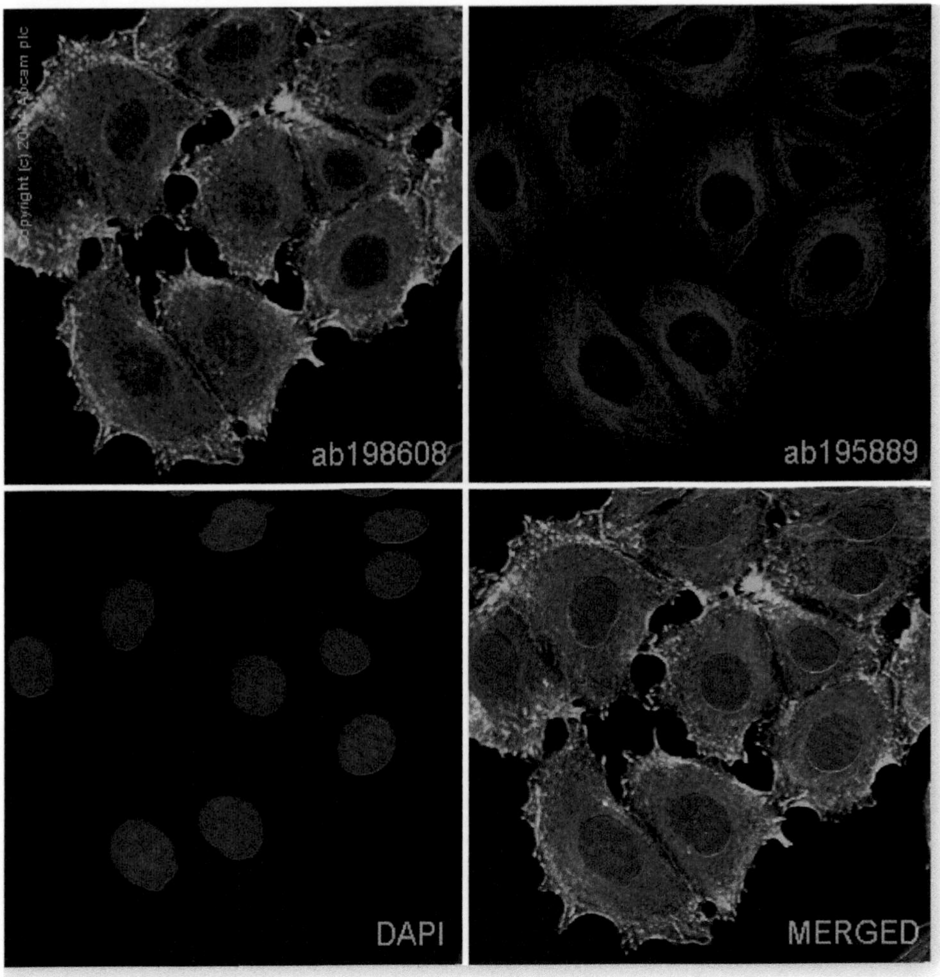

ab198608 staining alpha actinin 4 (microfilament marker) in MCF7 cells. The cells were fixed with 100% methanol (5 min), permeabilized with 0.1% Triton X-100 for 5 min and then blocked with 1% BSA/10% normal goat serum/0.3 M glycine in 0.1% PBS-Tween for 1 h. The cells were then incubated overnight at +4°C with ab198608 (Alexa Fluor® 488) at a 1/100 dilution (shown in green) and ab195889, mouse monoclonal to alpha tubulin (Alexa Fluor® 594), at a 1/250 dilution (shown in red). Nuclear DNA was labelled with DAPI (shown in blue). This product also gave a positive signal under the same testing conditions in MCF7 cells fixed with 4% formaldehyde (10 min)

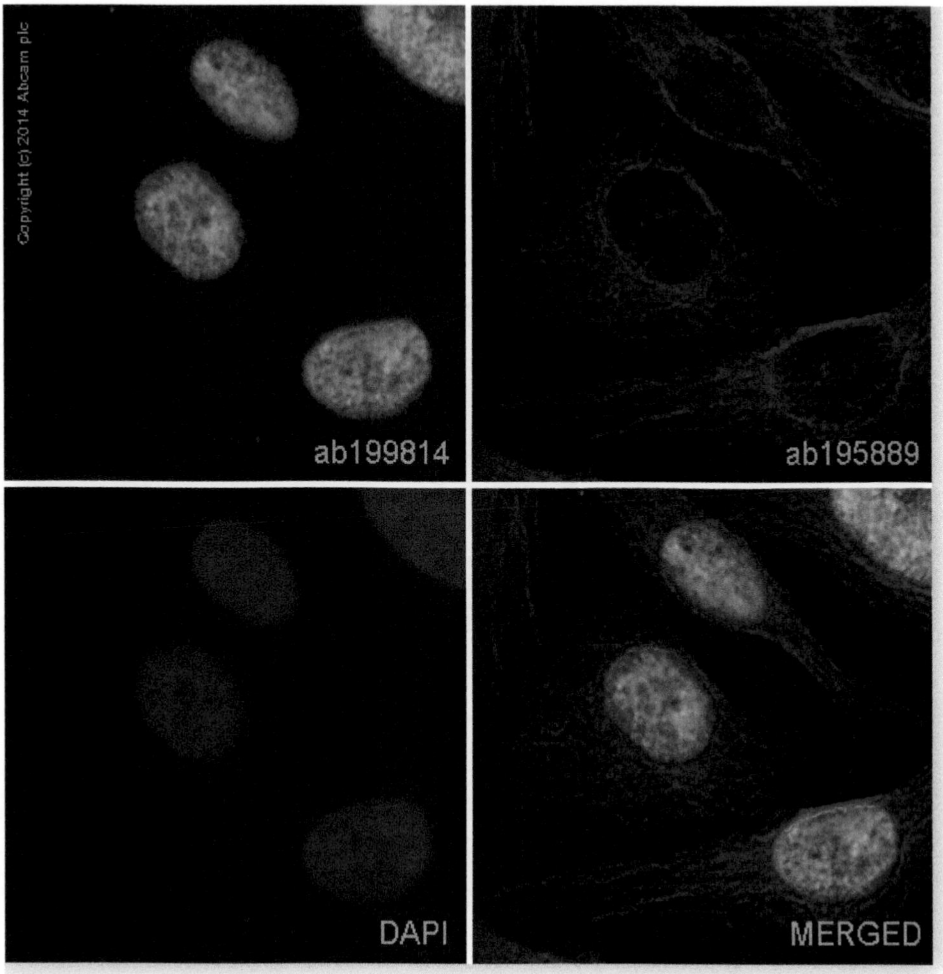

ab199814 staining YY1 (nuclear marker) in HeLa cells. The cells were fixed with 100% methanol (5 min), permeabilized with 0.1% Triton X-100 for 5 min and then blocked with 1% BSA/10% normal goat serum/0.3 M glycine in 0.1% PBS-Tween for 1 h. The cells were then incubated overnight at +4°C with ab199814 (Alexa Fluor® 488) at a 1/100 dilution (shown in green) and ab195889, mouse monoclonal to alpha tubulin (Alexa Fluor® 594), at a 1/250 dilution (shown in red). Nuclear DNA was labelled with DAPI (shown in blue). This product also gave a positive signal under the same testing conditions in HeLa cells fixed with 4% formaldehyde (10 min)

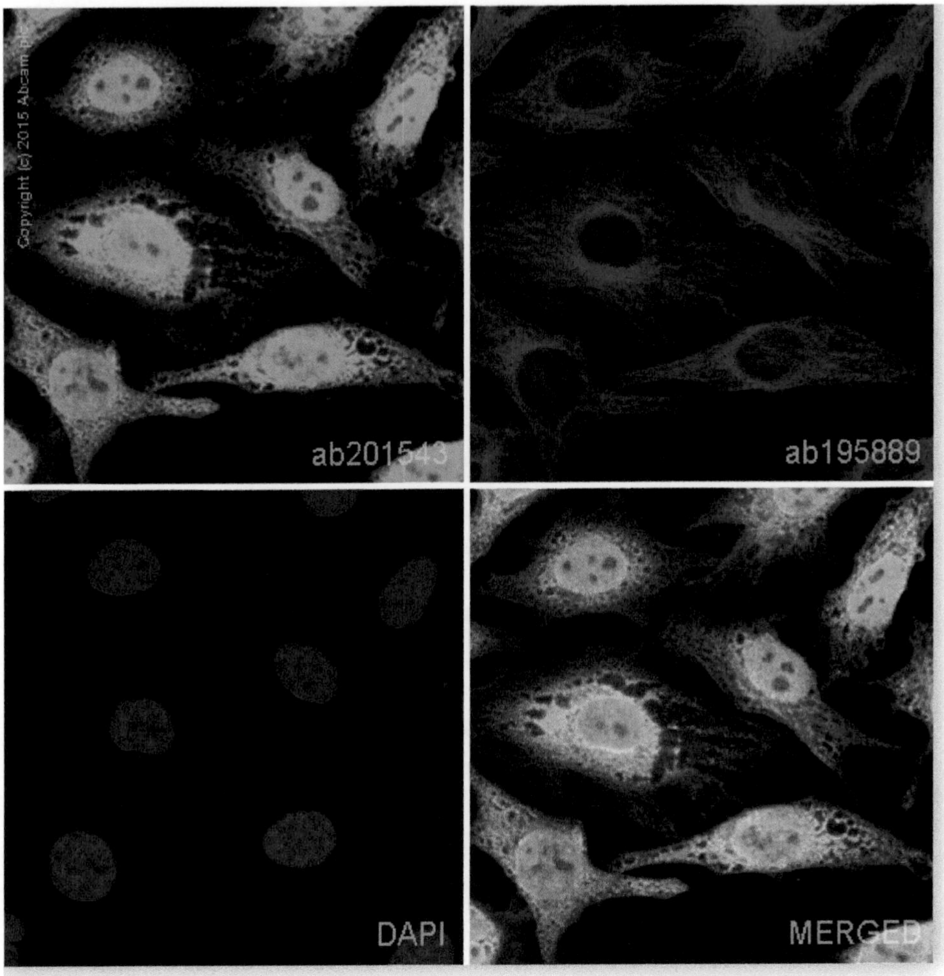

ab201543 staining SENP1 in HeLa cells. The cells were fixed with 4% formaldehyde (10 min), permeabilized with 0.1% Triton X-100 for 5 min and then blocked with 1% BSA/10% normal goat serum/0.3 M glycine in 0.1% PBS-Tween for 1 h. The cells were then incubated overnight at +4°C with ab201543 (Alexa Fluor® 488) at a 1/100 dilution (shown in green) and ab195889, mouse monoclonal to alpha tubulin (Alexa Fluor® 594), at a 1/250 dilution (shown in red). Nuclear DNA was labelled with DAPI (shown in blue). This product also gave a positive signal under the same testing conditions in HeLa cells fixed with 100% methanol (5 min)

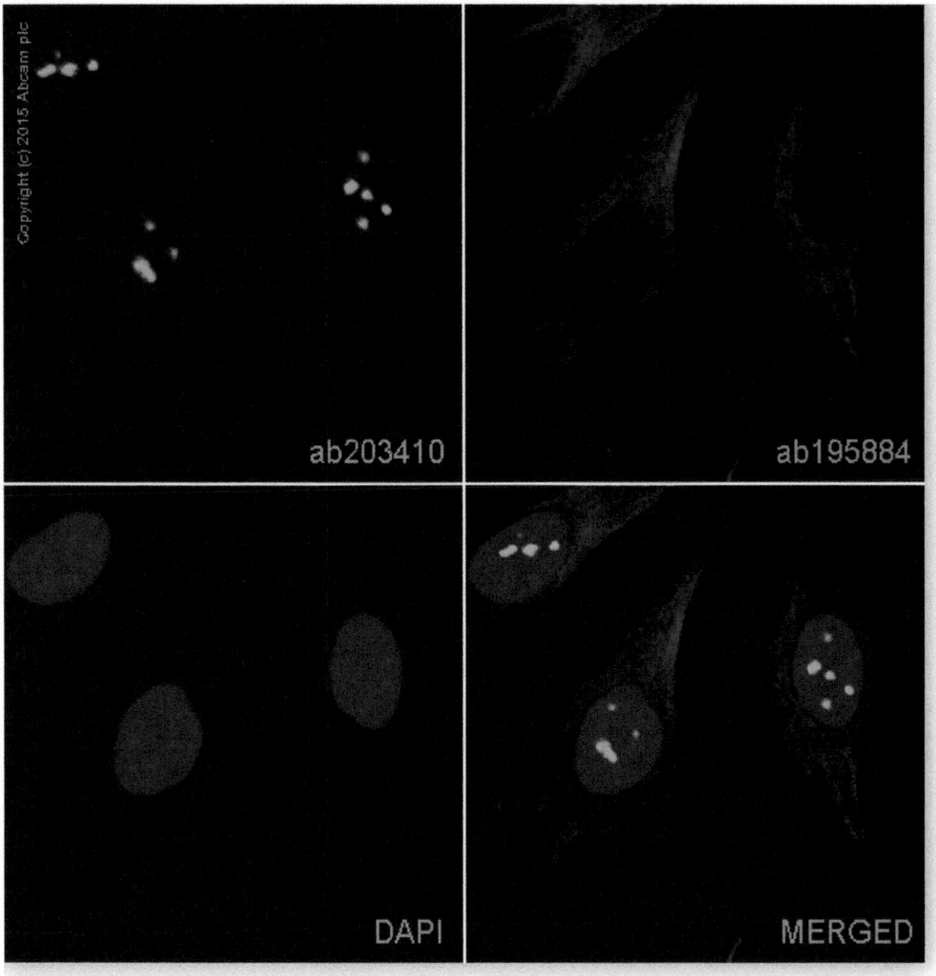

ab203410 staining fibrillarin (nucleolar marker) in HeLa cells. The cells were fixed with 100% methanol (5 min), permeabilized with 0.1% Triton X-100 for 5 min and then blocked with 1% BSA/10% normal goat serum/0.3 M glycine in 0.1%PBS-Tween for 1 h. The cells were then incubated overnight at +4°C with ab203410 (Alexa Fluor® 555) at 1/200 dilution (pseudo-coloured in green) and ab195884, rat monoclonal to tubulin (Alexa Fluor® 647), at 1/250 dilution (shown in red). Nuclear DNA was labelled with DAPI (shown in blue)

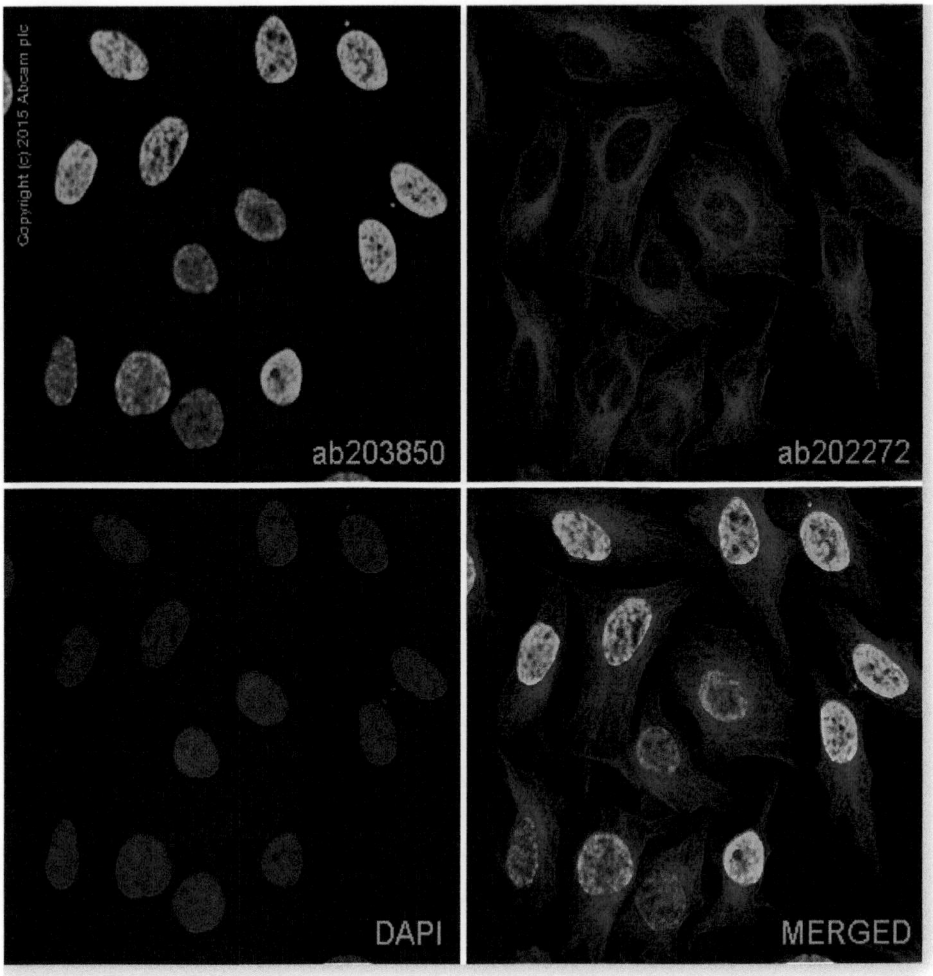

ab203850 staining histone H3 (dimethyl K9) in HeLa cells. The cells were fixed with 4% formaldehyde (10 min), permeabilized with 0.1% Triton X-100 for 5 min and then blocked with 1% BSA/10% normal goat serum/0.3 M glycine in 0.1% PBS-Tween for 1 h. The cells were then incubated overnight at +4°C with ab203850 (Alexa Fluor® 488) at a 1/100 dilution (shown in green) and ab202272, rabbit monoclonal to alpha tubulin (Alexa Fluor® 594), at a 1/250 dilution (shown in red). Nuclear DNA was labelled with DAPI (shown in blue). This product also gave a positive signal under the same testing conditions in HeLa cells fixed with 100% methanol (5 min)

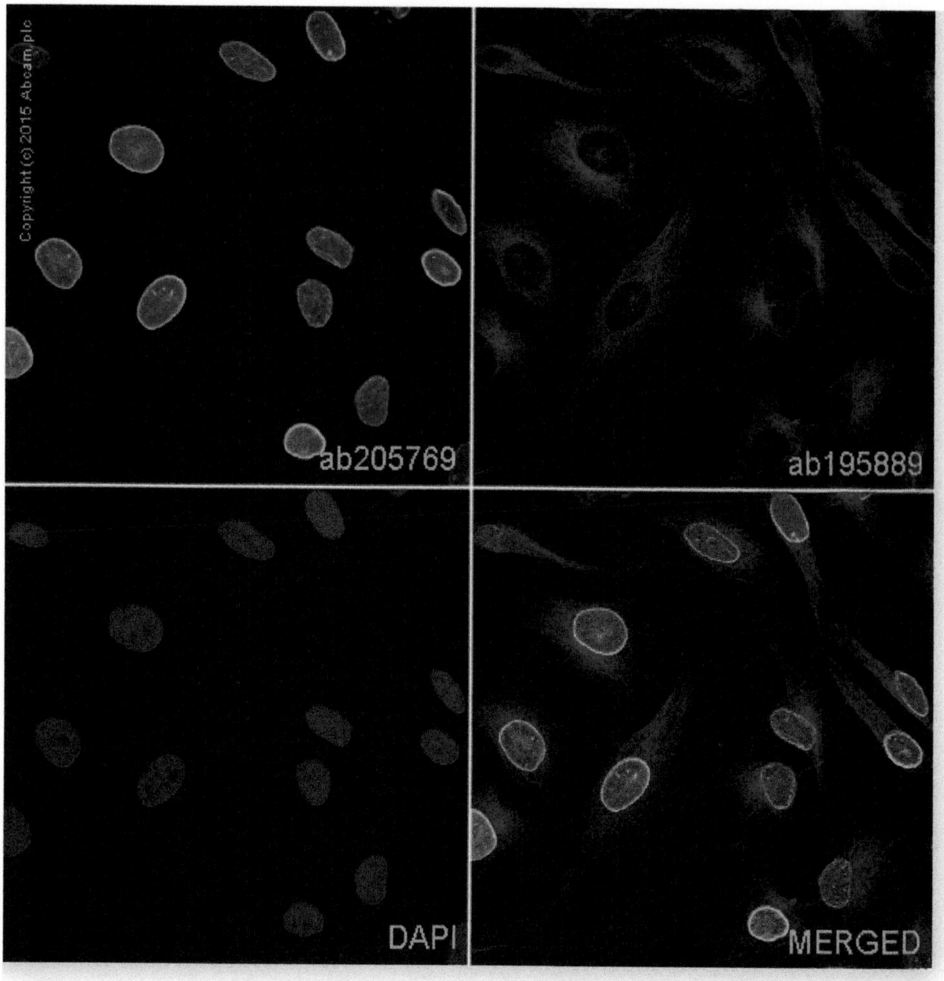

ab205769 staining lamin A+C (nuclear membrane marker) in HeLa cells. The cells were fixed with 4% formaldehyde (10min), permeabilized with 0.1% Triton X-100 for 5 min and then blocked with 1% BSA/10% normal goat serum/0.3 M glycine in 0.1% PBS-Tween for 1 h. The cells were then incubated overnight at +4°C with ab205769 (Alexa Fluor® 488) at 1/1000 dilution (shown in green) and ab195889, Mouse monoclonal [DM1A] to alpha tubulin (Alexa Fluor® 594), at 2 µg/mL (shown in red). Nuclear DNA was labelled with DAPI (shown in blue)

Immunohistochemistry

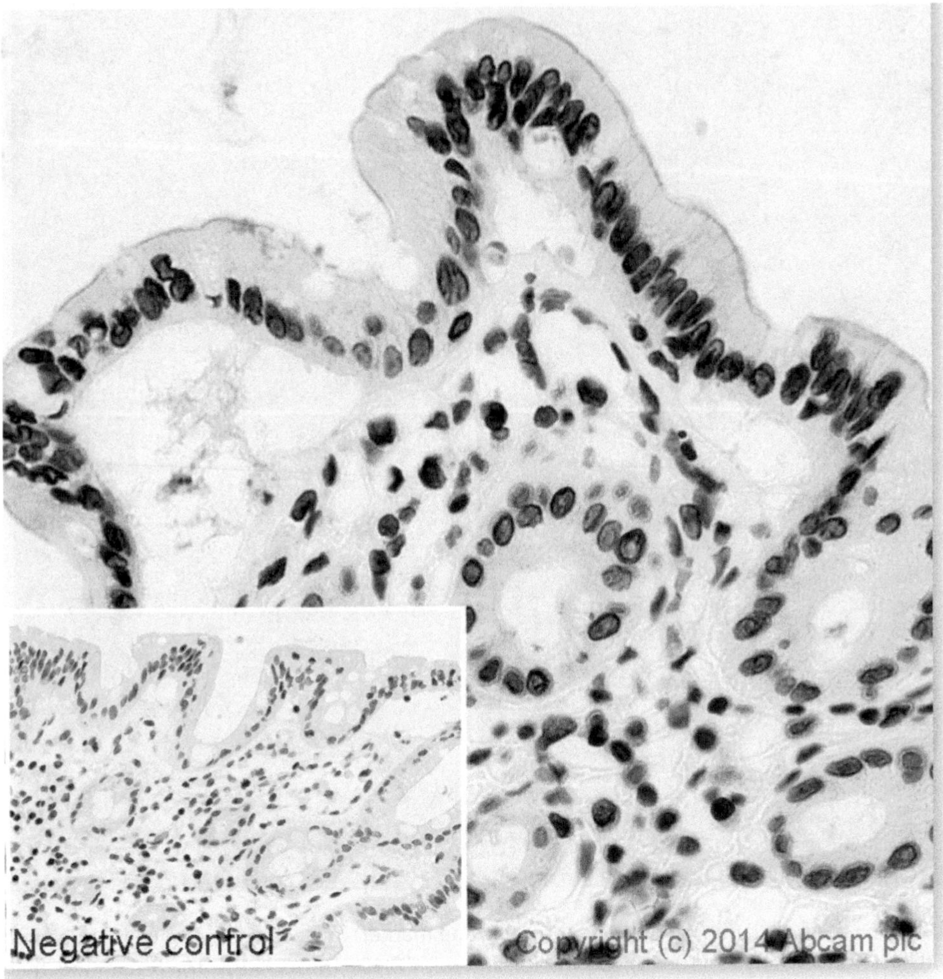

IHC image of lamin B1 (nuclear membrane marker) staining in a section of formalin-fixed paraffin-embedded normal human colon tissue*, performed on a Leica BOND™. The section was pre-treated using heat-mediated antigen retrieval with sodium citrate buffer (pH 6, epitope retrieval solution 1) for 20 min. The section was then incubated with ab194109 at 1/500 dilution for 15 min at room temperature. DAB was used as the chromogen. The section was then counterstained with haematoxylin and mounted with DPX. The inset negative control image is taken from an identical assay without primary antibody. **Tissue obtained from the Human Research Tissue Bank, supported by the NIHR Cambridge Biomedical Research Centre*

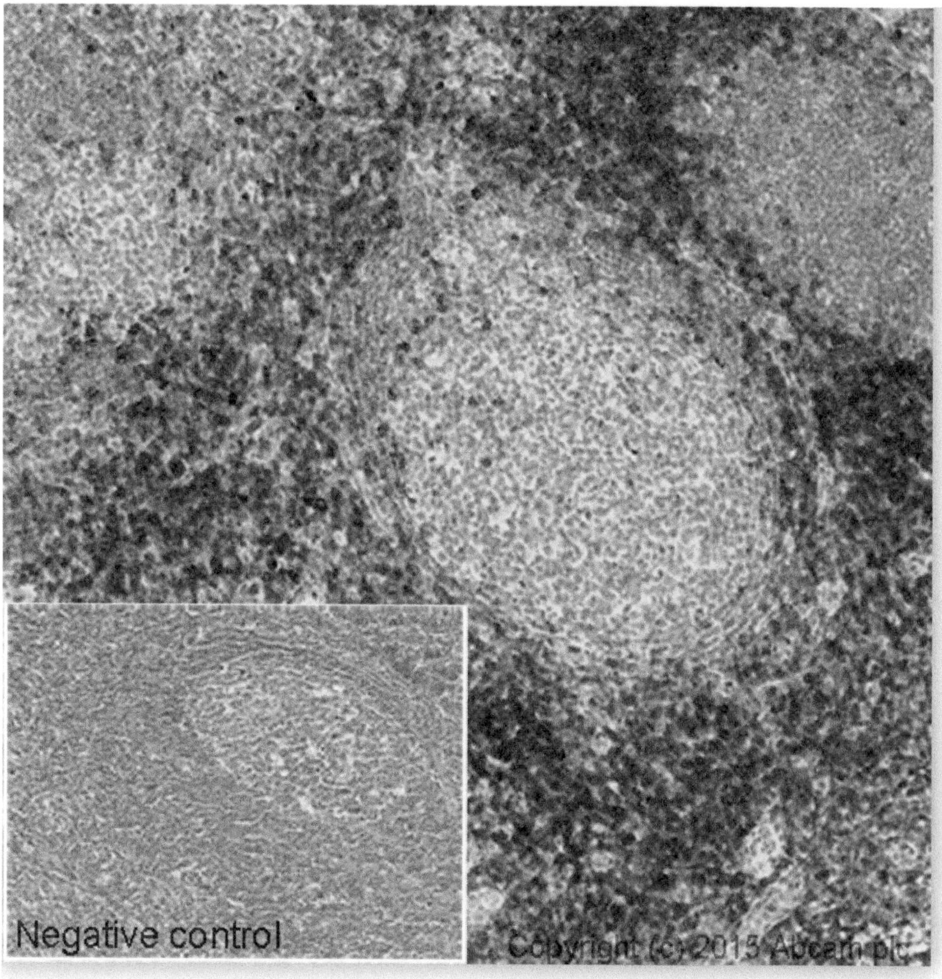

IHC image of CD7 (mature T-cell marker) staining in a section of formalin-fixed paraffin-embedded normal human tonsil*, performed on a Leica BOND™. The section was pre-treated using heat-mediated antigen retrieval with sodium citrate buffer (pH 6, epitope retrieval solution 1) for 20 min. The section was then incubated with ab199024 at 1/100 dilution for 15 min at room temperature. DAB was used as the chromogen. The section was then counterstained with haematoxylin and mounted with DPX. The inset negative control image is taken from an identical assay without primary antibody. **Tissue obtained from the Human Research Tissue Bank, supported by the NIHR Cambridge Biomedical Research Centre*

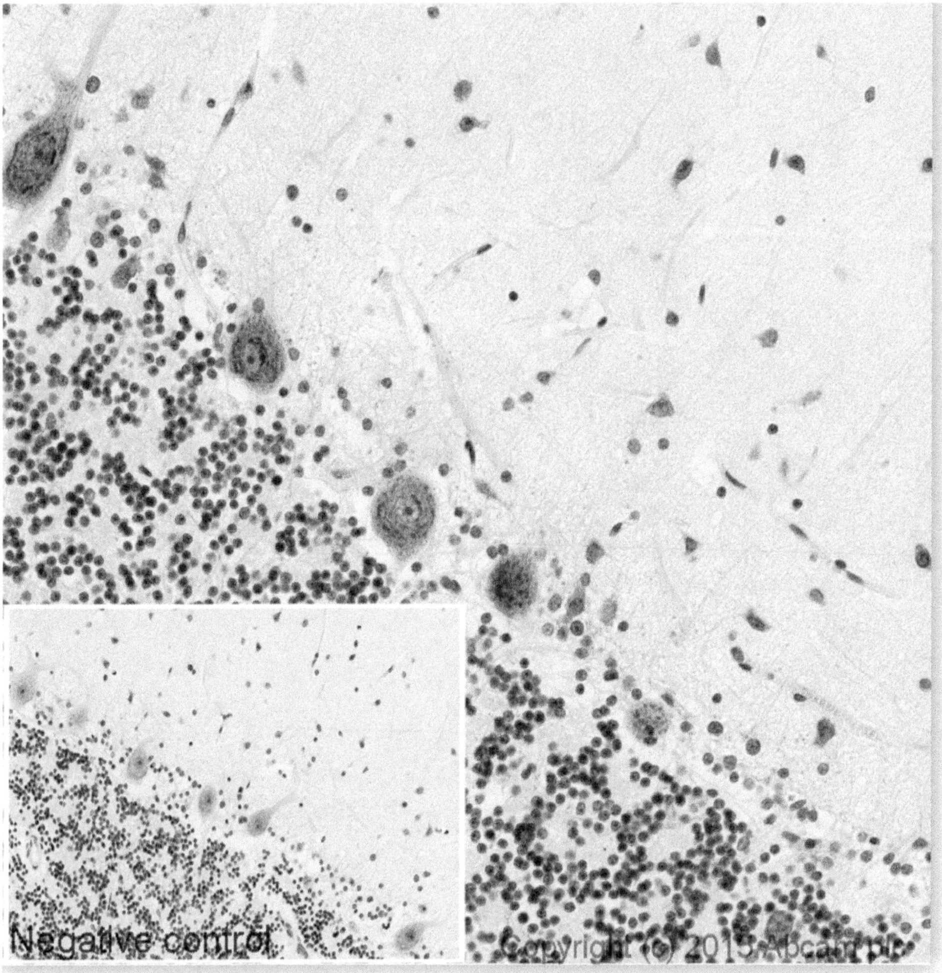

IHC image of adenosine A1 receptor staining in a section of formalin-fixed paraffin-embedded human cerebellum (normal), performed on a Leica BOND™. The section was pre-treated using heat-mediated antigen retrieval with sodium citrate buffer (pH 6, epitope retrieval solution 1) for 20 min. The section was then incubated with ab202551, 1/100 dilution, for 15 min at room temperature. DAB was used as the chromogen. The section was then counterstained with haematoxylin and mounted with DPX. The inset negative control image is taken from an identical assay without primary antibody

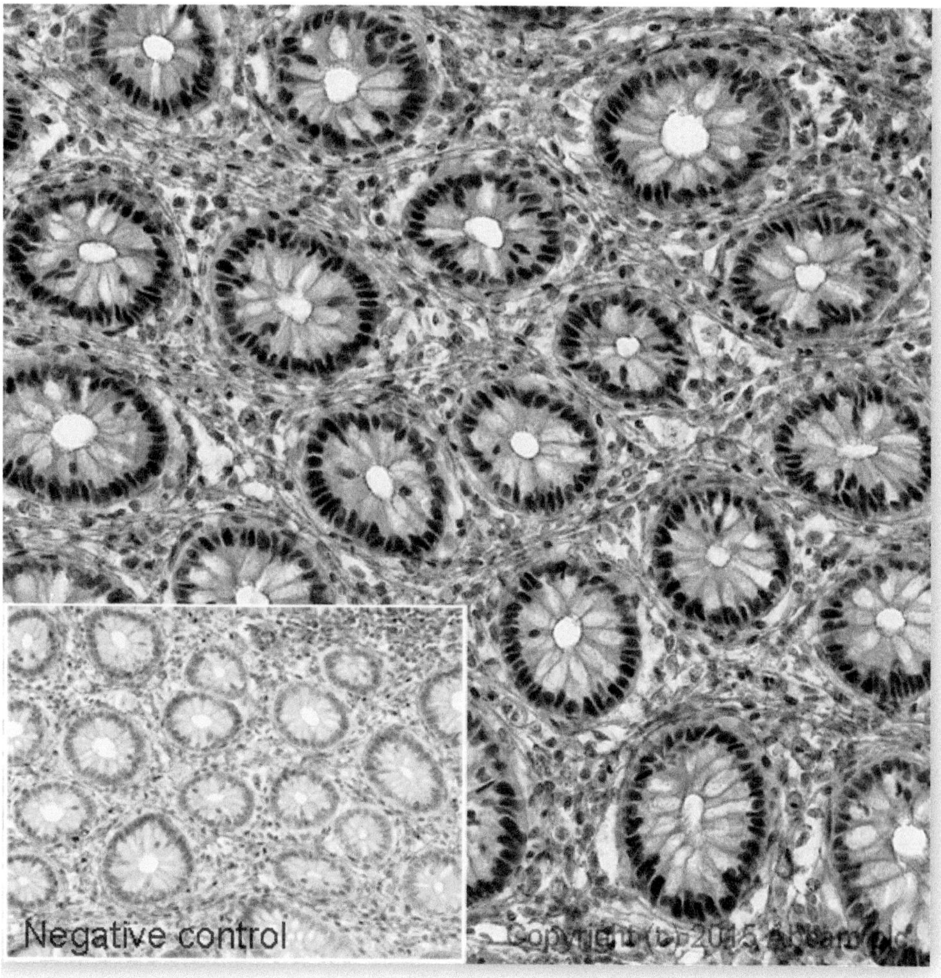

IHC image of MCM7 staining in a section of formalin-fixed paraffin-embedded human normal colon*. The section was pre-treated using pressure cooker heat-mediated antigen retrieval with sodium citrate buffer (pH 6) for 30 min and incubated overnight at +4°C with ab199498 at 1/100 dilution. DAB was used as the chromogen (ab103723), diluted 1/100 and incubated for 10 min at room temperature. The section was counterstained with haematoxylin and mounted with DPX. The inset negative control image is taken from an identical assay without primary antibody. **Tissue obtained from the Human Research Tissue Bank, supported by the NIHR Cambridge Biomedical Research Centre*

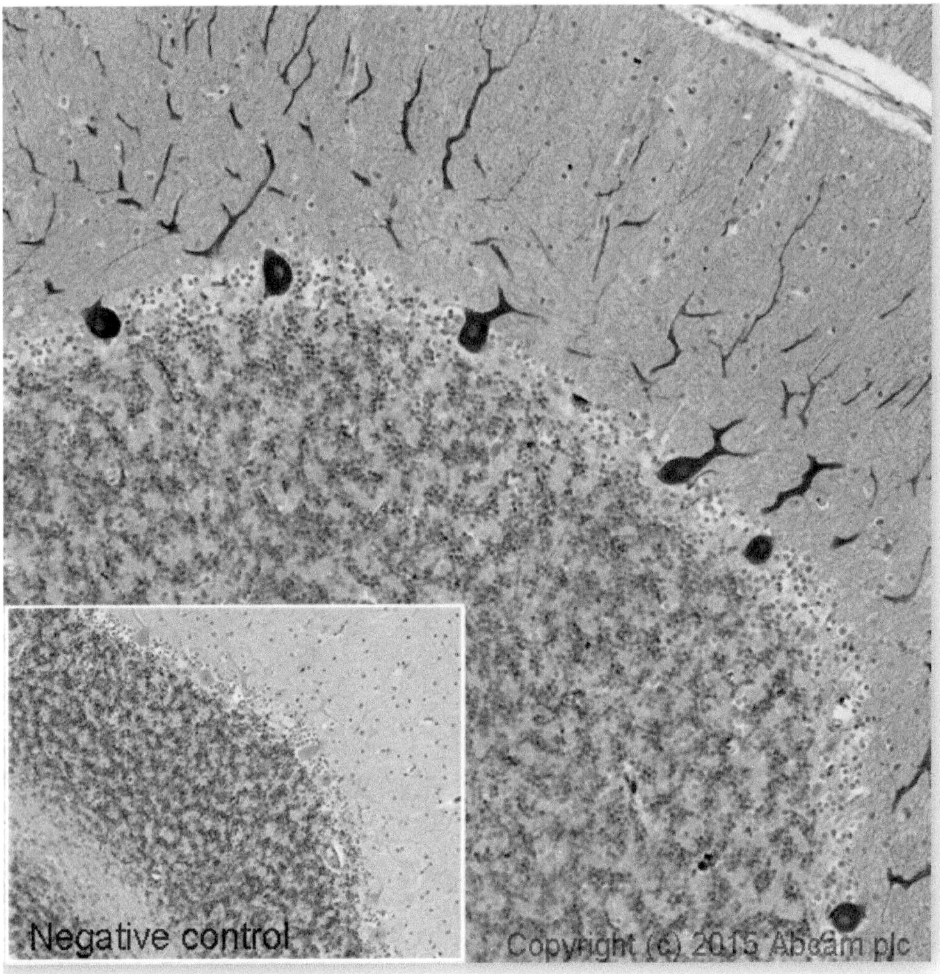

IHC image of LRRK2 staining in a section of formalin-fixed paraffin-embedded normal human cerebellum. The section was pre-treated using pressure cooker heat-mediated antigen retrieval with sodium citrate buffer (pH 6) for 30 min and incubated overnight at +4°C with ab195024 at 1/500 dilution. DAB was used as the chromogen (ab103723), diluted 1/100 and incubated for 10 min at room temperature. The section was counterstained with haematoxylin and mounted with DPX. The inset negative control image is taken from an identical assay without primary antibody

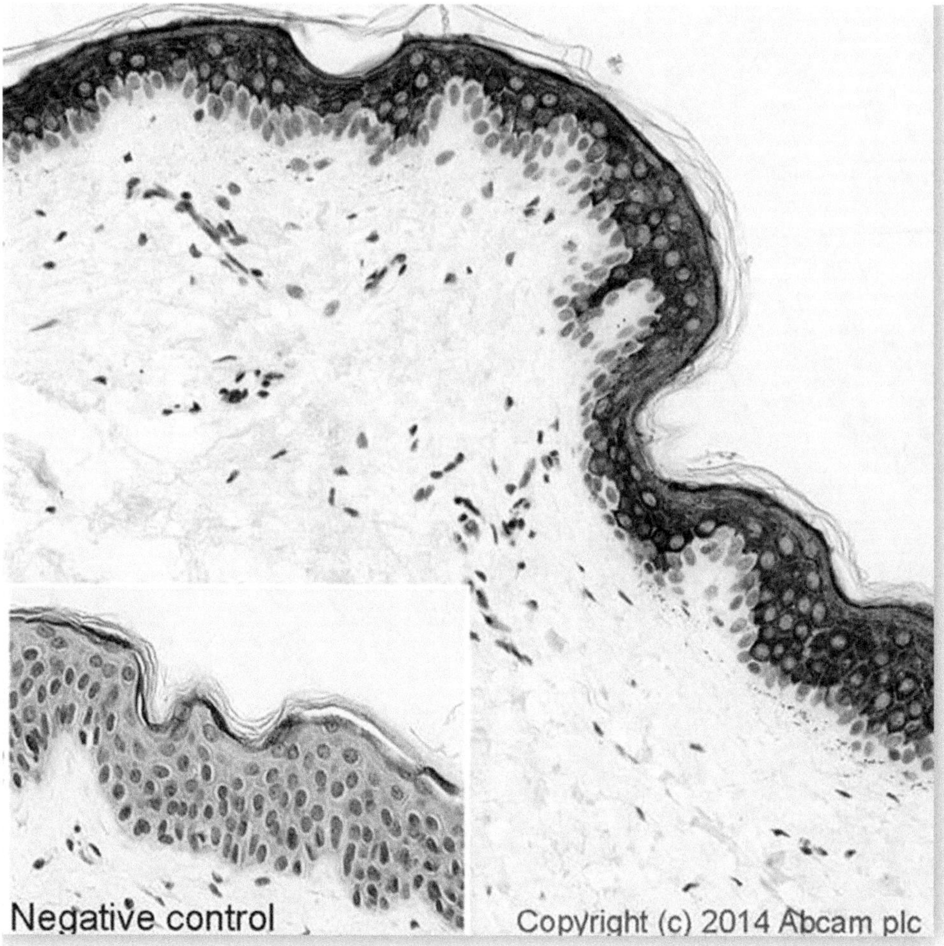

IHC image of cytokeratin 10 staining in a section of formalin-fixed paraffin-embedded normal human skin*, performed on a Leica BOND™. The section was pre-treated using heat-mediated antigen retrieval with sodium citrate buffer (pH 6, epitope retrieval solution 1) for 20 min. The section was then incubated with ab194232 at 1/100 dilution for 15 min at room temperature. DAB was used as the chromogen. The section was then counterstained with haematoxylin and mounted with DPX. The inset negative control image is taken from an identical assay without primary antibody. **Tissue obtained from the Human Research Tissue Bank, supported by the NIHR Cambridge Biomedical Research Centre*

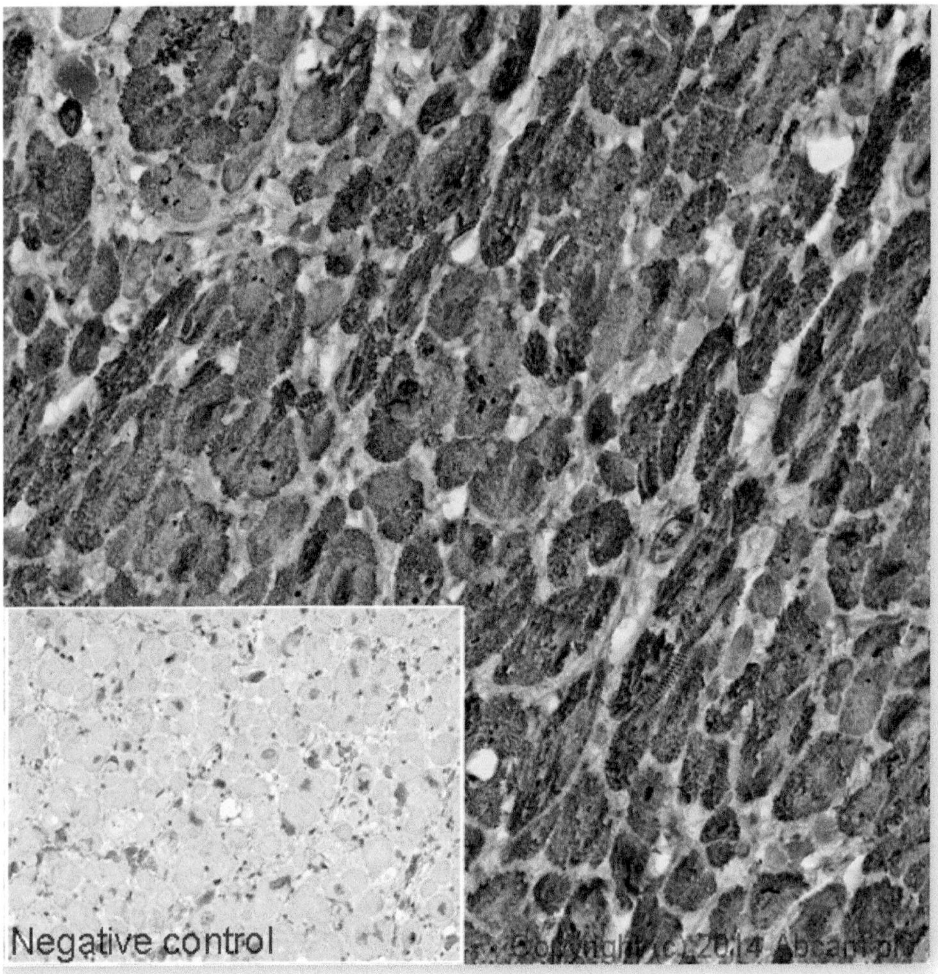

IHC image of cardiac troponin I staining in a section of formalin-fixed paraffin-embedded human heart muscle*. The section was pre-treated using pressure cooker heat-mediated antigen retrieval with sodium citrate buffer (pH 6) for 30 min and incubated overnight at +4°C with ab195529 at 1 μg/mL. DAB was used as the chromogen (ab103723), diluted 1/100 and incubated for 10 min at room temperature. The section was counterstained with haematoxylin and mounted with DPX. The inset negative control image is taken from an identical assay without primary antibody

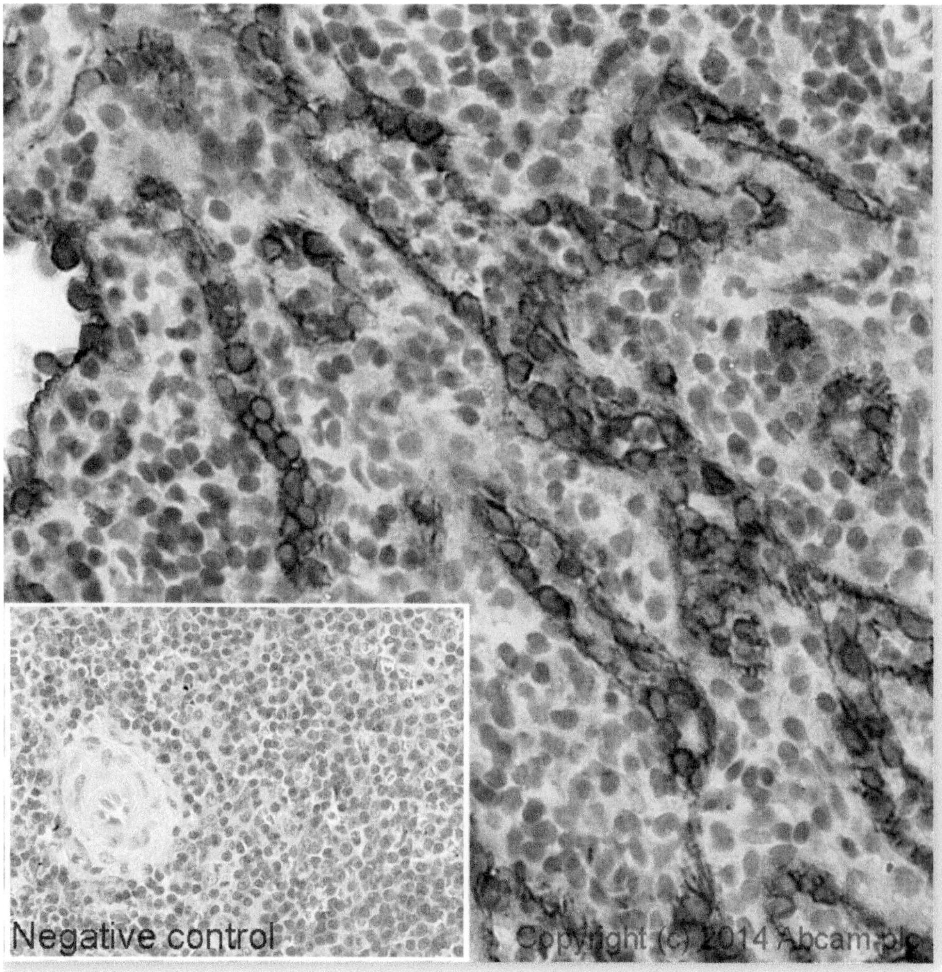

IHC image of VCAM1 staining in a section of formalin-fixed paraffin-embedded normal human spleen tissue*, performed on a Leica BOND™. The section was pre-treated using heat-mediated antigen retrieval with sodium citrate buffer (pH 6, epitope retrieval solution 1) for 20 min. The section was then incubated with ab195540 at 1/100 dilution for 15 min at room temperature. DAB was used as the chromogen. The section was then counterstained with haematoxylin and mounted with DPX. The inset negative control image is taken from an identical assay without primary antibody. **Tissue obtained from the Human Research Tissue Bank, supported by the NIHR Cambridge Biomedical Research Centre*

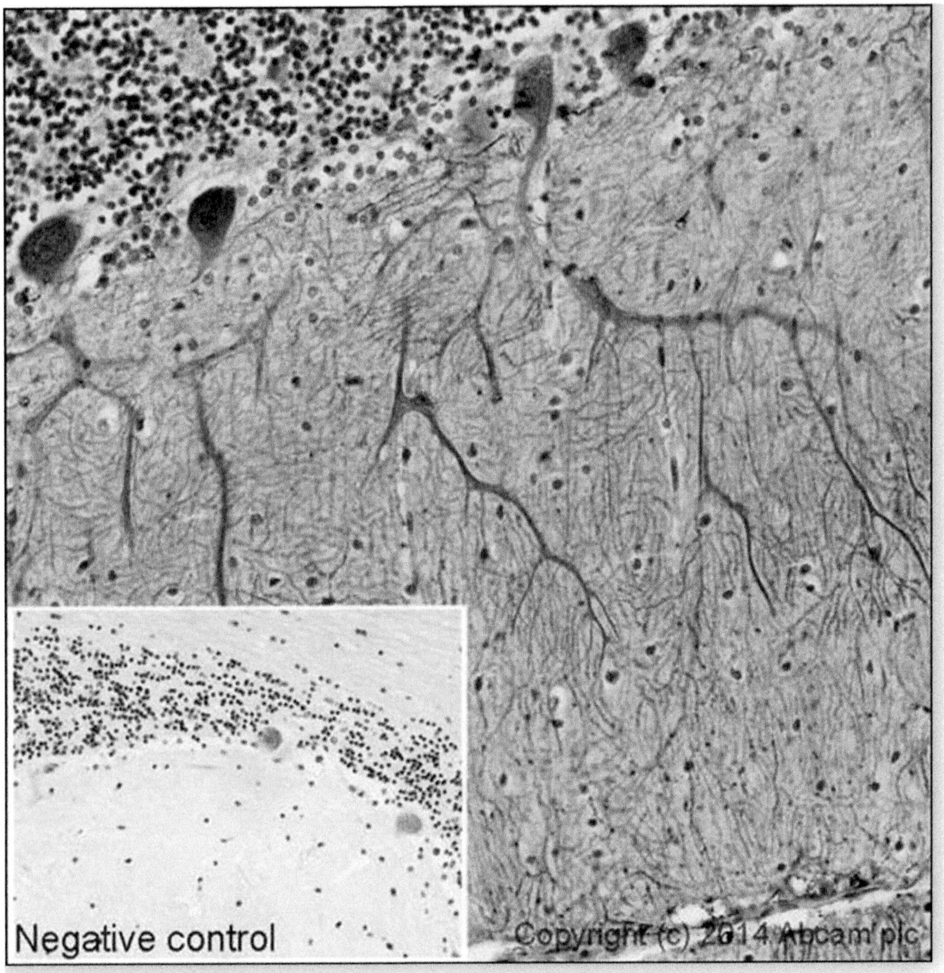

IHC image of metabotropic glutamate receptor 5 staining in a section of formalin-fixed paraffin-embedded normal human cerebellum, performed on a Leica BOND™. The section was pre-treated using heat-mediated antigen retrieval with sodium citrate buffer (pH 6, epitope retrieval solution 1) for 20 min. The section was then incubated with ab196482 at 1/100 dilution for 15 min at room temperature. DAB was used as the chromogen. The section was then counterstained with haematoxylin and mounted with DPX. The inset negative control image is taken from an identical assay without primary antibody

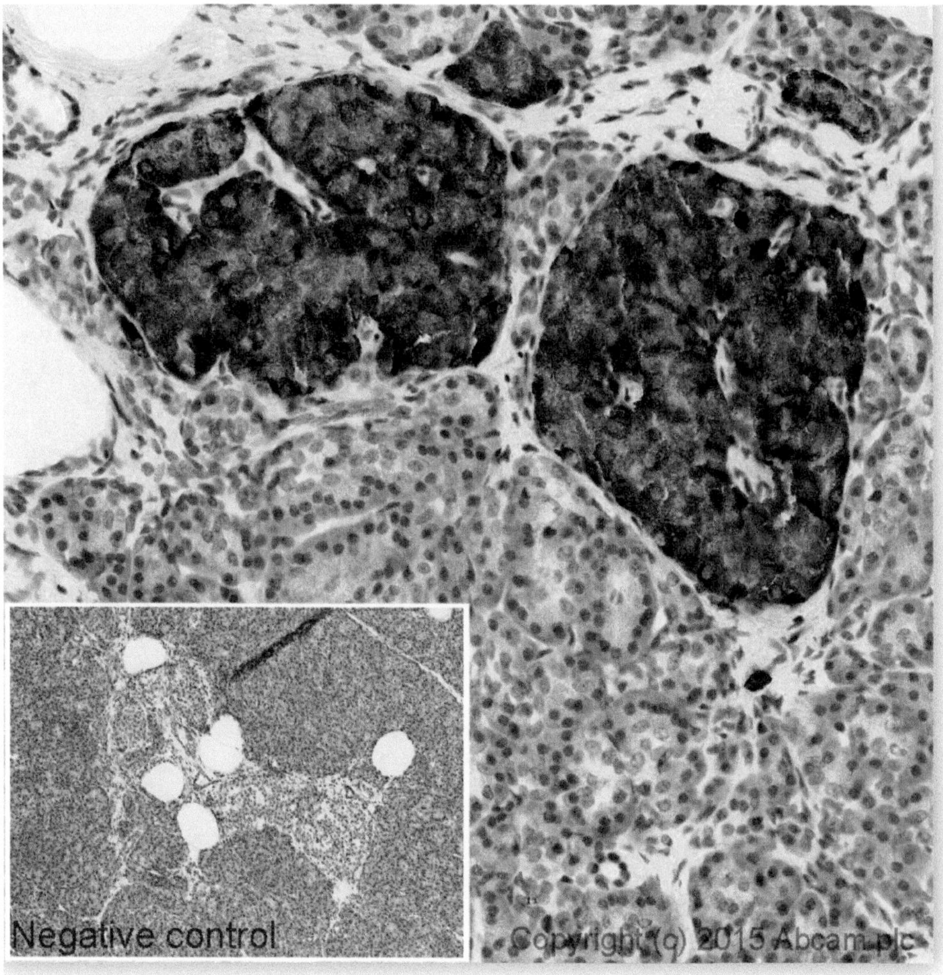

IHC image of chromogranin A staining in a section of formalin-fixed paraffin-embedded normal human pancreas*, performed on a Leica BOND™. The section was pre-treated using heat-mediated antigen retrieval with sodium citrate buffer (pH 6, epitope retrieval solution 1) for 20 min. The section was then incubated with ab199194, 1/1000 dilution, for 15 min at room temperature. DAB was used as the chromogen. The section was then counterstained with haematoxylin and mounted with DPX. The inset negative control image is taken from an identical assay without primary antibody. **Tissue obtained from the Human Research Tissue Bank, supported by the NIHR Cambridge Biomedical Research Centre*

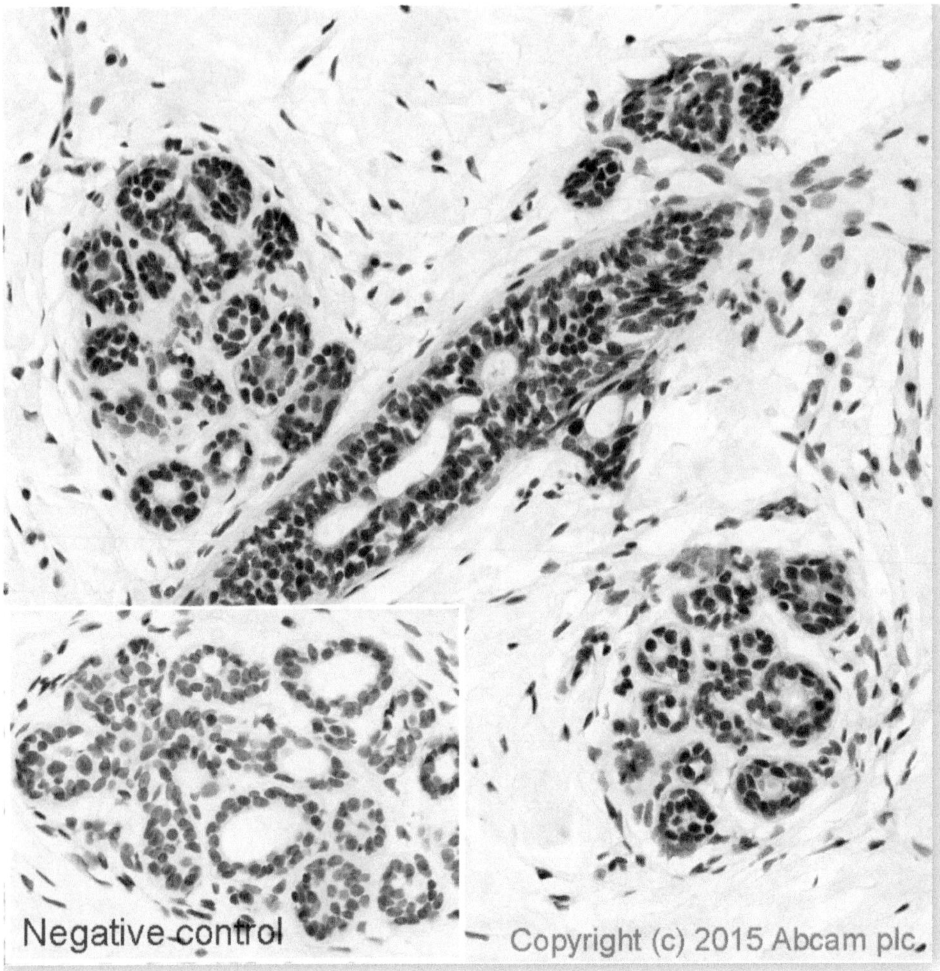

IHC image of PIM1 staining in a section of formalin-fixed paraffin-embedded normal human breast*, performed on a Leica BOND™. The section was pre-treated using heat-mediated antigen retrieval with sodium citrate buffer (pH 6, epitope retrieval solution 1) for 20 min. The section was then incubated with ab200889, 1/50 dilution, for 15 min at room temperature. DAB was used as the chromogen. The section was then counterstained with haematoxylin and mounted with DPX. The inset negative control image is taken from an identical assay without primary antibody. *Tissue obtained from the Human Research Tissue Bank, supported by the NIHR Cambridge Biomedical Research Centre*

ACKNOWLEDGEMENTS

The author gratefully acknowledges that the photomicrographic images shown in this chapter were produced using Abcam primary antibodies (www.abcam.com, with ab numbers referring to Abcam product catalogue numbers).

REFERENCES

1. Bancroft, J. and Stevens, A. (eds) (1996) *Theory and Practice of Histological Techniques*, 4th edn, Churchill Livingstone, NY, USA.
2. Monsan, P., Puzo, G. and Marzarguil, H. (1975) Etude du mecanisme d'etablissement des liaisons glutaraldehyde-proteines. *Biochimie*, **57**, 1281–1292.
3. Horobin, R.W. and Tomlinson, A. (1976) The influence of the embedding medium when staining for electron microscopy: the penetration of stains into plastic sections. *Journal of Microscopy*, **108**, 69–78.
4. Karnovsky, M.J. (1965) A formaldehyde-glutaraldehyde fixative of high osmolality for use in electron microscopy. *The Journal of Cell Biology*, **27**, 137A–138A.
5. Woods, A.E. and Ellis, R.C. (1994) *Laboratory Histopathology: A Complete Reference*, Churchill Livingstone, Edinburgh, UK.
6. Stead, R.H., Bacolini, M. and Leskovec, M. (1985) Update on the immunocytochemical identification of lymphocytes in tissue sections. *Canadian Journal of Medical Technology*, **47**, 162–178.
7. McLean, I.W. and Nakane, P.K. (1974) Periodate-lysine paraformaldehyde fixative. A new fixation for immunoelectron microscopy. *Journal of Histochemistry & Cytochemistry*, **22**, 1077–1083.
8. Holgate, C.S., Jackson, P., Pollard, K. *et al.* (1986) Effect fixation on T and B lymphocyte surface membrane antigen demonstration in paraffin-processed tissue. *The Journal of Pathology*, **149**, 293–300.
9. Miller RT (2001) Technical Immunohistochemistry: Achieving Reliability and Reproducibility of Immunostains. Society for Applied Immunohistochemistry Annual Meeting, 8 September, 2001, NY, USA.
10. Dorfman, D.M. and Shahsafaei, A. (1997) Usefulness of a new CD5 antibody for the diagnosis of T-cell and B-cell lymphoproliferative disorders in paraffin sections. *Modern Pathology*, **10**, 859–863.
11. Facchetti, F., Alebardi, O. and Vermi, W. (2000) Omit iodine and CD30 will shine: a simple technical procedure to demonstrate the the CD30 antigen on B5-fixed. *The American Journal of Surgical Pathology*, **24**, 320–322.
12. Abbodonzo, S.L., Allred, D.C., Lampkin, S. and Bank, P.M. (1991) Antigen retrieval immunohistochemistry review and future prospects in research and diagnosis over two decades. *Archives of Pathology & Laboratory Medicine*, **115**, 31–33.
13. Arnol, M.M., Srivastava, S., Fredenburgh, J. *et al.* (1996) Effects of fixation and tissue processing on immuohistochemical demonstration of specific antigens. *Biotechnic & Histochemistry*, **71**, 224–230.
14. Dapson, R.W. (1993) Fixation for the 1990's: a review of needs and accomplishments. *Biotechnic & Histochemistry*, **68**, 75–82.
15. Laurila, P., Virtanen, I., Wartiovaara, J. and Stenman, S. (1978) Fluorescent antibodies and lectins stain intracellular structure in fixed cells treated with nonionic detergents. *Journal of Histochemistry & Cytochemistry*, **26**, 251–257.
16. Hopwood, D. (1969) Fixatives and fixation: a review. *Histochemical Journal*, **1**, 323–360.
17. Bancroft, J. and Gamble, M. (eds) (2007) *Theory and Practice of Histological Techniques*, 6th edn, Elsevier Health Sciences, Oxford.

18. Kumar, G.L. and Rudbeck, L. (eds) (2009) *Dako Immunohistochemical Staining Methods*, 5th edn, Dako North America, Carpinteria.
19. DiVito, K.A., Charette, L.A., Rimm, D.L. and Camp, R.L. (2004) Long-term preservation of antigenicity on tissue microarrays. *Laboratory Investigation*, **84**, 1071–1078.
20. Manne, U., Myers, R.B., Srivastava, S. and Grizzle, W.E. (1997) Loss of tumor marker-immunostaining intensity on stored paraffin slides of breast cancer. *Journal of the National Cancer Institute*, **89**, 585–586.
21. Xie, R., Chung, J.-Y., Ylaya, K. *et al.* (2011) Factors influencing the degradation of archival formalin-fixed paraffin-embedded tissue sections. *Journal of Histochemistry & Cytochemistry*, **59**, 356.

CHAPTER FOUR

4

Multiple Immunochemical Staining Techniques

Sofia Koch

Abcam plc, Cambridge, UK

INTRODUCTION

Immunochemical staining of cells and tissues is a powerful technique that has emerged in cell biology to provide a direct visualization of proteins and a better understanding of the basic cellular functions. For this reason, it became an established tool for both research and diagnostic studies. With an increase in the complexity of the strategies used by researchers came the desire for multiple antigen visualization. This chapter describes the most common methods employed to successfully perform multiple immunochemical staining using brightfield, conventional epi-fluorescence and confocal microscopy.

Advantages and Technical Challenges

Multiple immunochemical staining allows the detection of two or more antigens within the same cell or tissue section, thus giving important simultaneous information on the

Immunohistochemistry and Immunocytochemistry: Essential Methods, Second Edition. Edited by Simon Renshaw.

cellular/sub-cellular location, biological function and relationships of various proteins [1, 2]. For instance, multiple immunochemical staining can be used:

- To investigate whether two or more antigens co-localize in a given subcellular location of the cell.
- To obtain information at the protein and DNA/mRNA levels. This can be done by combining *in situ* hybridization and immunochemical methods.
- To phenotype cells when no specific marker is available.
- To understand the distribution of a protein at the subcellular level by labelling antibodies against protein and a given organelle marker.

Overall, applying multiple immunochemical staining to samples saves time and reagents. For this reason, this technique is particularly important when the biological sample is scarce and there is a need to retrieve all possible information out of the available material.

Although the simultaneous staining of multiple antigens is seen as an attractive approach, researchers face various technical challenges that should be considered before starting any project:

- Elaborate protocols often have to be performed due to the limited number of possible combinations of primary antibodies. In fact, the majority of primary antibodies commercially available are raised in mouse or rabbit.
- The simultaneous use of fluorophores may lead to the overlap of their absorption and/or emission spectra. Sophisticated instrumentation and software can be used to resolve such spectral mixing issues. However, these are not available to some laboratories due to their expense and complexity. For this reason, it is essential that a combination of spectrally compatible fluorophores is considered beforehand.

Multiple Immunochemical Staining Method Selection

To ensure success, the researcher must carefully plan a multiple immunochemical staining project in advance. There are two main approaches to multiple immunochemical staining: immunofluorescence (IF), which is characterized by the use of fluorescent reporter labels; and immunoenzymatic, which is commonly performed using enzyme reporter labels. As described in Figure 4.1, the decision on which approach to follow will be determined by various factors such as the nature of the sample under investigation, the primary antibodies, the targets of interest and the time available to carry out the experiment. Importantly, the instrumentation available for the analysis of the results is also a crucial aspect to consider [2].

Both IF and immunoenzymatic staining can be performed on paraffin-fixed tissues, cell monolayers (such as 'Cytospin' and 'ThinPrep' preparations) and frozen sections. In addition, IF is the preferred approach when working with various adherent cells lines. It is easier to isolate the individual signals when using fluorescence, and cell lines often do not suffer from the same degree of auto-fluorescence as tissue sections because they possess

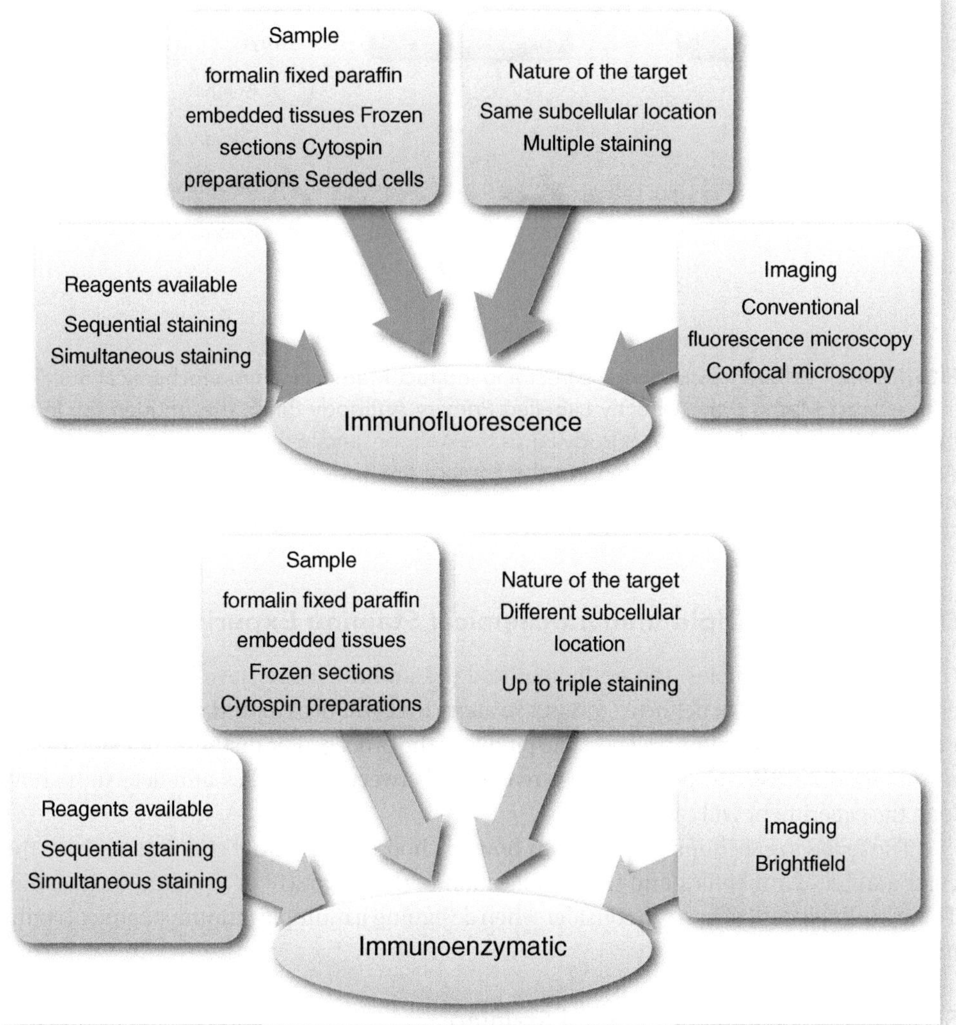

FIGURE 4.1 Aspects to be Considered When Selecting a Multi-Staining Method. The Choice to Perform Immunofluorescence or an Enzymatic-Based Assay will Depend on the Type of Sample, the Reagents, the Nature of the Target and the Imaging Tools Available

fewer components that naturally fluoresce, such as collagen (see p 144). The availability of reagents and the nature of the target are intrinsically associated as these will determine whether simultaneous staining or sequential staining is performed (as explained in 'Simultaneous Versus Sequential Staining'). Finally, the type of microscope available needs to be put into the equation. Brightfield microscopy is necessary to visualize immnoenzymatic staining. IF staining can be analysed using conventional epi-fluorescence or confocal microscopy, provided that the microscope has the necessary suitable excitation and emission hardware for the selected reporter labels (Fig. 4.2).

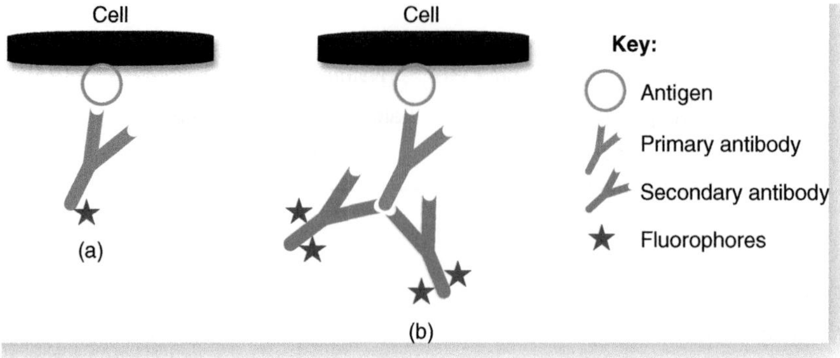

FIGURE 4.2 Differences between Direct and Indirect Method in Immunochemical Staining. In the Direct Method, the Directly Labelled Primary Antibody Binds the Antigen (a). In the Indirect Method, the Primary Antibody Binds the Antigen, and in the Second Step, Secondary Antibodies (Reporter Labelled) Bind to the Primary Antibody (b), Giving a Degree of Signal Amplification

Designing a Multiple Immunochemical Staining Experiment

Once the type of samples, the antibodies and the immunoassay have been decided, it is extremely important to define a staining strategy that will help to understand the interactions between the reagents and possible pitfalls of the process. For instance, by establishing a strategy, it is easier to prevent cross-reactivity between antibodies and determine how long the experiment will take.

The following sections will focus on how to choose the optimal staining method, the combination of antibodies and the visualization of the target (IF or immuno-enzymatic). These are important aspects to consider when designing a multiple immunochemical staining experiment.

Simultaneous versus Sequential Staining

When the labelling of several antigens is being done with antibodies raised in different species, then simultaneous staining is the preferred method. This approach is significantly quicker when compared to sequential staining because the reagents can be mixed together. Sequential immunostaining is the preferred method when primary antibodies are raised in the same species and are of the same isotype. Adopting a sequential method significantly increases the time required for staining antigens but, in contrast, prevents cross-reactivity-related problems [1]. The decision of whether to perform simultaneous or sequential staining is straightforward when researchers want to stain their samples using IF techniques. However, if an enzyme-based method is desired, one has to consider the nature of the antigens apart from the availability of primary antibodies. In fact, immune-enzymatic techniques should only be done when the antigens are located in different subcellular locations or even different cells. This is because the excessive deposit of coloured chromogen/enzyme precipitant end product can cover the antigen and prevent it being accessed by another primary antibody [2]. Therefore, in such cases, IF has to be performed.

Direct versus Indirect Method

Before starting an immunochemical staining experiment, the decision must be made whether to perform a direct or an indirect method depending on the available primary antibodies and the expression levels of the antigens [3]. A direct method of detection uses primary antibodies that are directly conjugated to a reporter label. Consequently, there is only one step of incubation that makes the staining procedure straightforward. In contrast, the indirect method of detection consists of two incubation steps: one for the primary antibody and the other for the secondary reporter labelled antibody. Although more complex and time consuming, this method is more sensitive as several secondary antibody molecules are likely to bind a single primary antibody molecule; hence, the number of reporter labels in the vicinity of the primary antibody (and therefore antigen) is increased, giving rise to a stronger signal [4]. This process is known as 'signal amplification' (see p 62). Importantly, it is possible to combine both methods (directly reporter labelled and unlabelled primary antibodies) if this helps the researcher avoiding species cross-reactivity during a multiple immunochemical staining experiment.

Choosing the Primary and Secondary Antibody Combination

A multiplex experiment, which uses primary antibodies raised in distinct host species, is probably the most simple and straightforward way of performing multiple immunochemical staining. Nevertheless, it is still important to spend time considering the most appropriate set of secondary antibodies and blocking serum for the combination of primary antibodies previously decided [5]. Figure 4.3 describes the most frequent combinations of primary antibodies that one can select depending on the available reagents.

Mainly, the researcher needs to consider using [5]:

- Primary antibodies raised in different host species (unless they are directly conjugated to reporter labels).
- Secondary antibodies raised against the respective host species of the primary antibodies, but both raised in the same species, so that normal serum from the host species of the secondary antibodies can be used to in the blocking buffer (see p 64).
- Secondary antibodies that have been pre-adsorbed against immunoglobulins from (i) the species of the biological sample being tested and (ii) the species(s) used to raise the other secondary and primary antibody combination in order to reduce the risk of unwanted cross-reactivity.

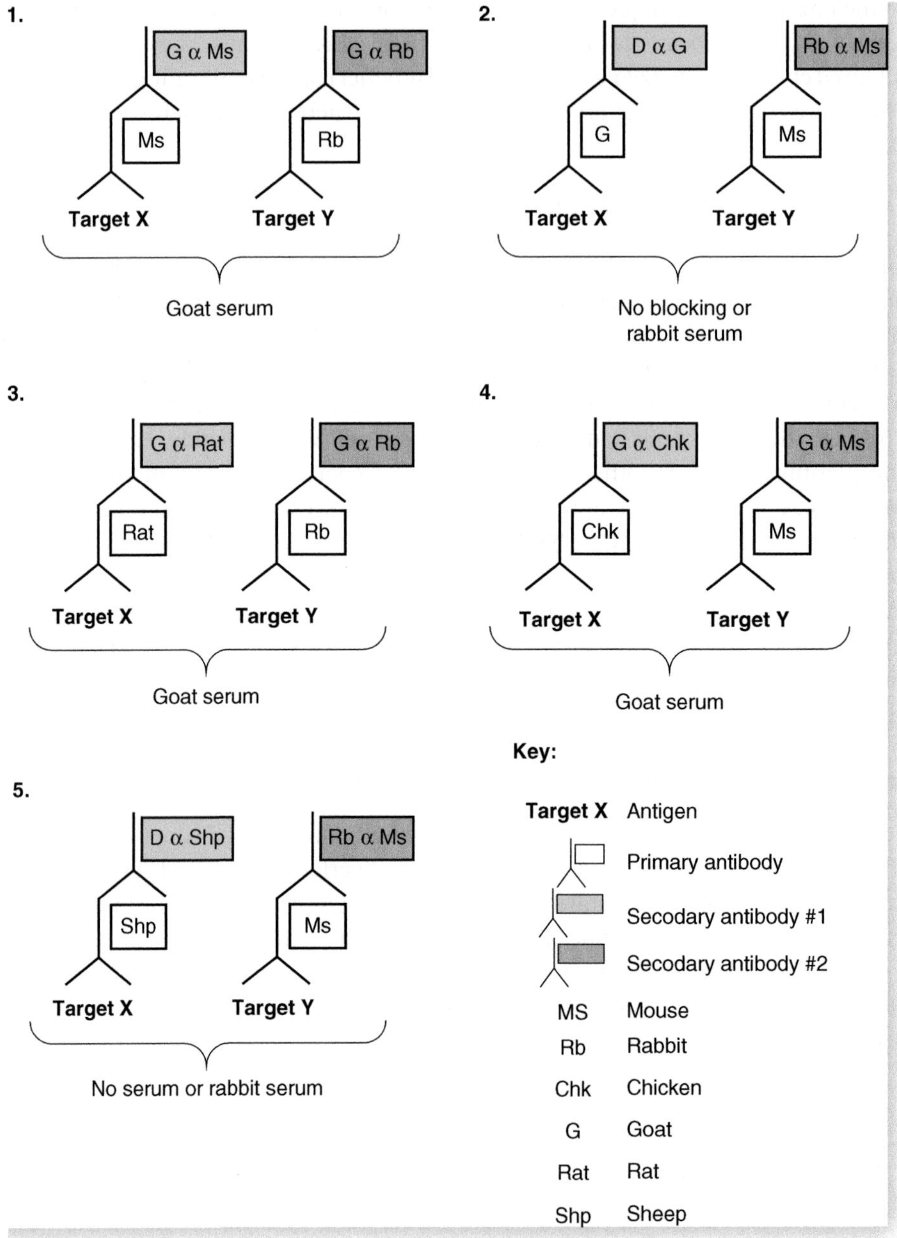

FIGURE 4.3 Combinations of Primary and Secondary Antibodies to Be Used to Prevent Cross-Reactivity

Choosing between Enzyme and Fluorescent Reporter Labels

Inevitably, when performing a multiple immunochemical staining experiment, one has to decide on how to label the targets of interest. As previously stated, the detection of antigens can either be done by using antibodies conjugated to fluorescent or enzymatic reporter labels. A discussion of whether to use fluorescent or enzymatic reporter labels depending on whether or not you are using paraffin embedded tissue sections, frozen tissue sections or cytological preparations can be found p 62.

Enzyme— Like fluorescence, chromogen detection also allows for the visualization of multiple antigens. However, as explained earlier, chromogens are often used when antigens are confined to distinct locations in the cells or tissue, because the overlap of different chromogen/enzyme precipitates (antigen co-localisation) can make results interpretation difficult. Careful consideration should therefore be chosen as to the colour combinations. Red/brown colour combinations (such as HRP/DAB; and AP/New Fuchsin) are not considered suitable where co-localization of the signal is to be defined by eye because the colour mix seen when co-localization occurs is undefinable. Red/blue (such as AP/Fast Blue BB; and HRP/AEC) or turquoise/red (such as ß-GAL/X-gal; and AP/Fast Red TR) are much more definable by eye, creating a purple–brown and purple–blue colour mix, respectively. Performing immunochemical staining using each primary antibody in isolation (one per identical test specimen, as well as all together) can help to isolate the individual staining patterns. However, an easier and more elegant way of distinguishing co-localized signals is by using a modern microscope software package with 'spectral un-mixing' capabilities. This has the ability to digitally isolate the different coloured signals from the multiple chromogen/enzyme precipitates, all from a single test specimen containing all of the primary antibodies used in the multiple immunochemical staining experiment.

Additionally, Fast Blue BB and X-gal are quite diffuse and inefficient chromogens when compared to AEC and Fast Red TR, so these are best reserved (when possible) for the most abundant of antigens, reserving AEC and Fast Red TR for the least abundant. There may also be a need to re-optimize the titres for primary antibodies when visualizing them using these chromogens to take this effect into account.

When using HRP/DAB combination, it is always worth nearing in mind the DAB 'shielding' (or 'sheltering') effect. The brown/black precipitate generated as the end product of this reaction can essentially cover the immunoreagents used to create the signal, thus preventing the subsequent binding of other antibodies. This can be very advantageous when carrying out sequential multiple immunochemical staining using primary antibodies raised in the same species because it can prevent the binding of the second secondary antibody (which is intended to only target the second primary antibody) to the first primary antibody. The concentration of the first antibody must be carefully titrated, since if too high, the DAB shielding may not go to completion, thus allowing the second secondary to bind, leading to unwanted staining. DAB shielding may also be a problem where two antigens are closely co-localized, preventing detection of the second primary with the second secondary antibody. DAB is therefore generally unsuitable for multiple immunochemical staining when co-localization of antigens is expected.

A final (very important) consideration is to ensure that mounting media used is compatible with both enzyme/chromogen end products (see p 74).

Fluorescence— When designing a multiple immunochemical staining experiment using fluorescent reporter labels (fluorochromes), it is imperative to first model the excitation and emission characteristics of the fluorochromes and counterstains being used, to ensure that there is no spectral overlap. This is best done using many of the free online fluorescence spectral analyser (viewer) programmes. Such programmes are excellent tools. Not only do they display the absorption and emission spectra of various fluorophores in comparison to each other, but they often allow the user to enter their specific filter-set characteristics, to see exactly the absorption and emission wavelengths.

Spectral overlap can lead to a phenomenon known as 'bleed-through' or 'cross-talk', characterized by detecting a fluorochrome's emission in the filter set of another fluorochrome. Typically, when two fluorochromes are spectrally similar, the emission of a fluorophore of a shorter excitation wavelength will be more prevalent in the emission profile of one that has a longer excitation wavelength.

Using combinations of fluorochromes with narrow emission spectra that emit in different spectral areas (e.g. a blue, green, orange and far red) help to prevent this problem. Microscope filters should also be chosen wisely to closely match the absorption and emission spectra of the reporter labels being used. Fortunately, since many filters come in a cube as a 'set', the manufacturer has already done this for us. Theoretical models aside, it is always pertinent to carry out each staining set in isolation (as well as together in the actual multiple immunochemical fluorescent staining experiment) and view them with the opposing filter set(s), in order to verify that there is indeed no cross-talk. However, such cross-talk can be rectified using the same spectral un-mixing capabilities as described earlier for co-localized chromogen/enzyme precipitates. In such cases, viewing each staining set in isolation is a critical part of the un-mixing process.

When designing a multiple immunochemical staining experiment, one should always try to use the brightest fluorochrome (the one with the highest quantum yield, see p 32) to visualize the least abundant antigen and to avoid the quenching of weaker signals. Conversely, the dimmest fluorochrome (the one with the lowest quantum yield) should therefore be used for the most abundant antigen. Similarly, detection system offering the highest degree of signal amplification should be reserved for the least abundant antigen. This is not always practically possible, however, and parameters such as antibody concentrations and microscope exposure settings should always be carefully optimized to reduce any such negative effects.

Determine the Experimental Controls

The use of antibodies in tissues or cells sometimes gives unexpected staining patterns, which are not related to the specific binding of the primary antibodies to the correct

TABLE 4.1 Controls Required for a Multiple Immunochemical Staining Experiment, Using Two Primary Antibodies (Ab1 and Ab2), and Their Corresponding Secondary Antibodies (S1 and S2), Respectively

Control Specimen	Reagent Combination	Purpose
1	Ab1 and S1	To assess each primary antibody staining pattern in isolation and check for bleed-through when viewed using the opposing filter set
2	Ab2 and S2	
3	Ab1, S1 and S2	To ensure that the secondaries do not cross-react with each other
4	Ab2, S2 and S1	
5	Ab1 and S2	To ensure that the secondaries do not cross-react with the other primary antibody
6	Ab2 and S1	
7	S1	To assess general non-specific binding of the secondary
8	S2	
9	No antibodies	To assess auto-fluorescence in each wavelength used

antigens. For this reason, the researcher should carefully set up various controls before initiating any experiment [6]. These controls will ensure that all aspects of the experiment are carried out correctly and will guarantee accuracy and reliability. The use of positive and negative controls (for both antigens and reagents) is covered in detail in Chapter 5, p 136. However, additional controls need to be used when performing multiple immunochemical staining [6]. Table 4.1 shows the controls required for a multiple immunochemical staining experiment, using two primary antibodies (Ab1 and Ab2), and their corresponding secondary antibodies (S1 and S2), respectively:

Antigen Retrieval

Antigen retrieval is often essential for the demonstration of antigens. Incorrect antigen retrieval can lead to unwanted background and/or staining patterns, as can over-retrieval. It is therefore imperative to ensure that the antigen retrieval technique(s) used does not negatively affect the other antigen. Ideally, one antigen retrieval technique should be used, even if it means that one antigen is not optimally retrieved, but retrieved well enough for it to be adequately detected. The first step is to ascertain the optimal antigen retrieval conditions for both of the antigens. The best case scenario is that the optimal retrieval conditions for each antigen will be the same. If not, retrieving both antigens separately with each technique will demonstrate how each is affected, and the one with the best compromise can be identified.

Heat-mediated antigen retrieval can be used to effect when performing multiple immunochemical staining because subsequent rounds of it remove previous antibodies, leaving behind just the enzyme/chromogen precipitate. Therefore, multiple staining patterns can be built up in a sequential manner (a round of heat-mediated antigen retrieval followed by immunochemical staining), with no issues of reagent cross-reactivity. However, multiple rounds of antigen retrieval can negatively alter the staining patterns of certain antigens, so the order of antigen staining must be carefully planned. For instance, if a multiple immunochemical staining experiment uses three separate primary antibodies, the staining pattern of each must be assessed after one, two and three rounds of antigen retrieval, so see if the expression pattern changes (and if the tissue and enzyme/chromogen precipitates survive, for that matter.) The staining pattern of each antibody is then observed after each round of antigen retrieval to ascertain the optimal order of immunochemical staining.

Double Staining Using Same-Species Primary Antibodies

Multiple immunochemical staining can be a straightforward method when the primary antibodies are raised in different species. However, many antibodies used in research are mouse monoclonals, suggesting that working with same-species primary antibodies is almost inevitable at some point. The biggest limitation of working with antibodies raised in the same host species is cross-reactivity, with the secondary antibodies unable to distinguish between the different target–primary antibody complexes. Despite being a challenging technique, several double-staining protocols have now been successfully established.

Directly Reporter Labelled Primary Antibodies

As briefly explained in 'Direct versus Indirect Method' section, directly reporter-labelled primary antibodies can be used in direct methods of immunostaining. In the context of a double staining using same-species primary antibodies, the researcher can use directly reporter-labelled antibodies not only to save time but to also prevent cross-reactivity and non-specific binding of the secondary antibodies. This method lacks sensitivity when one is interested in visualizing low-expressed targets because there is no signal amplification step provided by the secondary antibody [3]. As previously mentioned, when designing a multiple immunochemical staining experiment, one should always try to use the brightest fluorochrome (one with the highest quantum yield, see p 32) in conjunction with the least sensitive signal amplification method to visualize the most abundant antigen and to avoid the quenching of weaker signals. For this reason, conjugated primary antibodies are preferably used when the target is expressed in abundance.

Class and Subclass-Specific Secondary Antibodies

Although raised in the same host species, antibodies might present different classes of immunoglobulins, such as IgG or IgM, or different subclasses of IgG, such as mouse IgG1 and mouse IgG2. In these cases, it is simpler and effective to use isotype-specific secondary antibodies, which match the class or subclass of the primary antibodies used. However, caution should be exercised to ensure that the isotype-specific secondary antibodies are indeed specific! This can easily be assessed by performing immunochemical staining with each primary antibody in isolation, but with the incorrect secondary, to see if there is any signal (the desired outcome being no signal).

Using Reporter Label Conjugated Fab Fragments as Blocking Agents When Using Primary Antibodies Raised in the Same Species (see Protocol 5)

Fab fragments are monovalent, which means they have a single antigen-binding site. These are frequently used to block endogenous immunoglobulins in tissue sections or on cell surfaces. They can also be used to bind to and effectively block the surface of primary immunoglobulins, thus preventing any additional antibodies from binding. This effectively allows double immunochemical staining to be performed using primary antibodies raised in the same species. The example shown in Figure 4.4 considers two IgG mouse monoclonal primary antibodies. Following the incubation of the sample with the first primary antibody (step 1, which binds to antigen 'Y'), a polyclonal reporter label conjugated anti-mouse IgG Fab fragment is introduced (step 2), which binds to the surface of the primary antibody, completely blocking any available binding sites along its heavy and light chains. Since they are monovalent molecules, the Fab fragments will be unable to bind any additional mouse primary antibodies that are added to the experiment. At this point, the second primary antibody (also mouse IgG) is applied and allowed to bind to antigen 'X'. This is then subsequently detected with another polyclonal anti-mouse IgG secondary antibody, which can be either Fab or whole molecule in nature (step 3), conjugated to a reporter label that is different from that of the initial Fab fragments in step 2. To reiterate, this additional secondary antibody will not be able to bind to the first primary antibody introduced in step 1, since all of the available antibody binding sites on the initial primary antibody are already occupied by the Fab fragments.

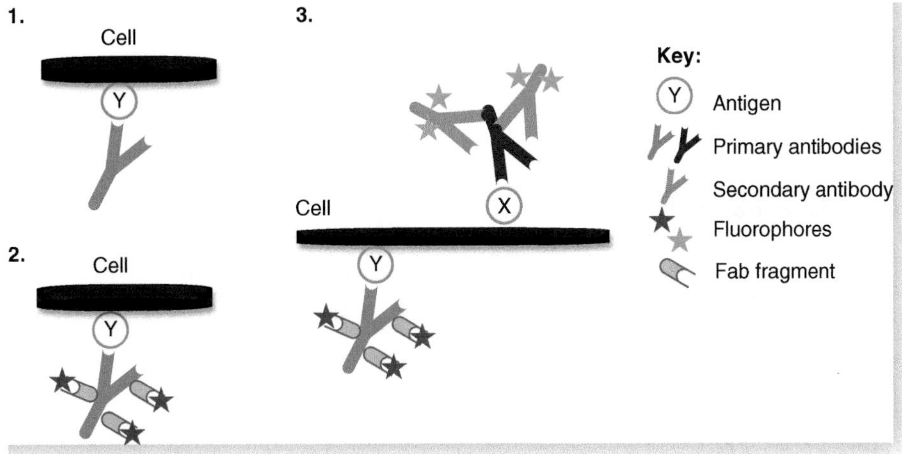

FIGURE 4.4 Double Immunostaining Using Conjugated Fab Fragments

Using Unconjugated Fab Fragments to Effectively Change the Species in Which a Primary Antibody is Raised (see Protocol 6)

Fab fragments can also bind to a primary antibody, effectively presenting it as a different species.

The example shown in Figure 4.5 considers two IgG mouse monoclonal primary antibodies. Following incubation of the sample with the initial primary antibody (step 1, which binds to antigen 'Y'), a goat polyclonal anti-mouse IgG Fab fragment is introduced (step 2), which binds to the surface of the primary antibody, completely blocking any available binding sites along its heavy and light chains. This also has the effect of changing the raised species (identity) of the initial primary antibody from mouse to goat. Aside, since they are monovalent molecules, the Fab fragments will be unable to bind any additional mouse primary antibodies that are added to the experiment. The initial primary antibody can then be detected using a polyclonal reporter label conjugated rabbit anti-goat IgG (H+L)

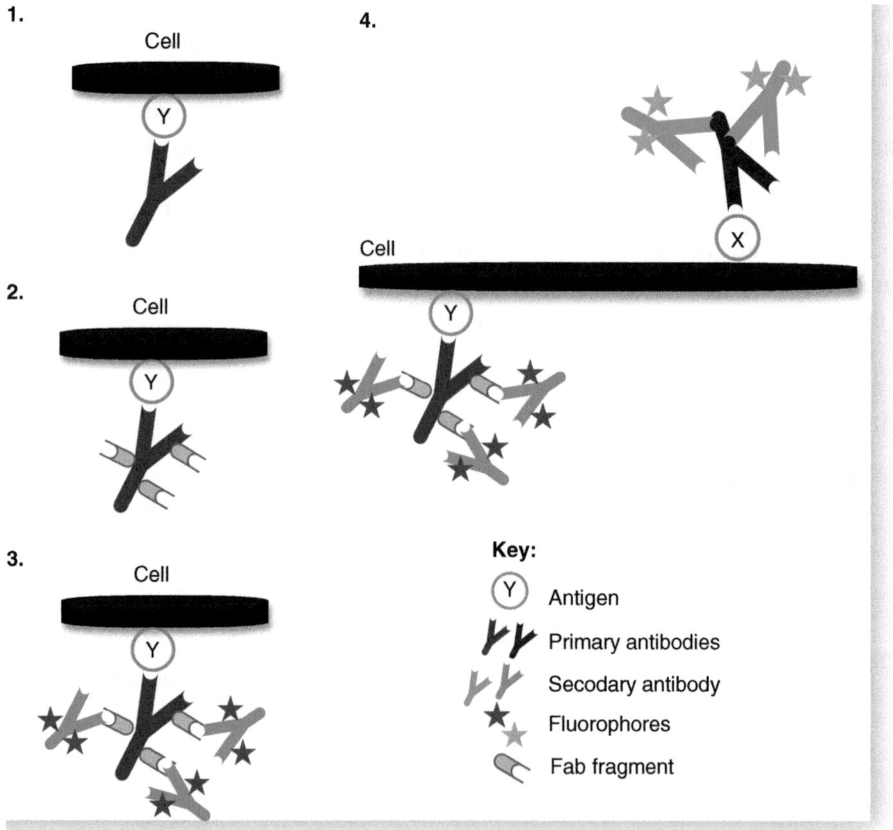

FIGURE 4.5 Double Immunostaining Using Unconjugated Fab Fragments

or anti-$F(ab')_2$ secondary antibody (step 3). It is imperative that this secondary antibody (effectively used in this instance as a tertiary antibody) must not recognize the host species of the primary antibodies or the secondary antibody which will be used to bind the second primary antibody. It is therefore ideal to have both secondary antibodies raised in the same species (for example, rabbit)

At this point, the second primary antibody (also mouse IgG) is applied and allowed to bind to antigen 'X'. This is then subsequently detected with a polyclonal anti-mouse IgG secondary antibody, which can be either Fab or whole molecule in nature (step 4), conjugated to a reporter label that is different from that of the rabbit anti-goat antibody used in step 3.

To reiterate, the secondary antibody used in step 4 will not be able to bind to the initial primary antibody introduced in step 1 because all of the available antibody-binding sites on the initial primary antibody are already occupied by the Fab fragments, effectively changing its identity from mouse to goat.

METHODS AND APPROACHES

Multiple reporter label immunochemical staining techniques are considered as advanced immunochemical staining techniques. Indeed, the protocols used are only slightly modified versions of single reporter label immunochemical staining protocols. The majority of the steps, reagents, principles and theories are the same. Similarly, the protocols can be further customized as long as the considerations discussed earlier in this chapter and in chapter 3 are observed. The reader is therefore advised to familiarize themselves with the theory and practice of single reporter label immunochemical staining protocols before attempting to perform multiple.

Protocol 1 – Simultaneous fluorescence: directly conjugated primary antibodies raised in same species

- ***Primary antibodies:***
  - ***2×, raised against different antigens***
  - ***raised in the same species***
  - ***added simultaneously***
  - ***directly conjugated, each to a different fluorescent reporter label (e.g. Alexa Fluor® 488 and Alexa Fluor® 594)***
- ***Secondary antibodies: none***

1. Culture and fix cells optimally (see p 37)
2. Rinse cells for 3 × 5 min in PBS containing 0.1% Tween-20 (see p 65)
3. Permeabilize cells using PBS containing 0.1%Triton X-100 for 10 min at room temperature (see p 65)
4. Block non-specific protein–protein interactions with PBS containing 0.1% Tween-20, 1% BSA and 0.3 mM glycine for 30 min at room temperature (see p 65)

5. Incubate cells with the optimally diluted directly reporter label-conjugated antibodies, made up in PBS containing 0.1% Tween-20 and 1% BSA for 1 h at room temperature, or overnight, at 4 °C (see p 66)
6. Rinse cells for 3 × 5 min in PBS containing 0.1% Tween-20
7. Incubate cells with suitable counterstain(s), made up according to the manufacturer's instructions (see p 67)
8. Rinse cells for 3 × 5 min in PBS containing 0.1% Tween-20
9. Mount in a suitable mounting media and analyse by fluorescence microscopy (see p 67)

Protocol 2 – Simultaneous fluorescence: unconjugated primary antibodies raised in different species

- ***Primary antibodies:***
  - ***2×, raised against different antigens***
  - ***raised in different species (e.g. mouse and rabbit)***
  - ***added simultaneously***
  - ***unconjugated***
- ***Secondary antibodies:***
  - ***raised in the same species (e.g. goat)***
  - ***added simultaneously***
  - ***directly conjugated, each to a different fluorescent reporter label (e.g. Alexa Fluor® 488 and Alexa Fluor® 594)***
  - ***each raised against the species of one of the primary antibodies***

1. Culture and fix cells optimally (see p 37)
2. Rinse cells for 3 × 5 min in PBS containing 0.1% Tween-20 (see p 65)
3. Permeabilize cells using PBS containing 0.1%Triton X-100 for 10 min at room temperature (see p 65)
4. Block non-specific protein–protein interactions with PBS containing 0.1% Tween-20, 1% BSA and 0.3 mM glycine for 30 min at room temperature (see p 65)
5. Incubate cells with the optimally diluted unconjugated antibodies, made up in PBS containing 0.1% Tween-20 and 1% BSA for 1 h at room temperature, or overnight, at 4 °C (see p 66)
6. Rinse cells for 3 × 5 min in PBS containing 0.1% Tween-20
7. Incubate cells with the optimally diluted directly reporter label-conjugated secondary antibodies, made up in PBS containing 0.1% Tween-20 and 1% BSA for 30 min at room temperature (see p 66)
8. Rinse cells for 3 × 5 min in PBS containing 0.1% Tween-20
9. Incubate cells with suitable counterstain(s), made up according to the manufacturer's instructions (see p 67)
10. Rinse cells for 3 × 5 min in PBS containing 0.1% Tween-20
11. Mount in a suitable mounting media and analyse by fluorescence microscopy (see p 67)

Protocol 3 – Sequential enzymatic: unconjugated primary antibodies raised in different species

- ***Primary antibodies:***
  - ***2×, raised against different antigens***
  - ***raised in different species (e.g. mouse and rabbit)***
  - ***added sequentially***
  - ***unconjugated***
- ***Secondary antibodies:***
  - ***one directly conjugated to AP (raised against the species of the first primary antibody)***
  - ***one directly conjugated to biotin (raised against the species of the second primary antibody)***
  - ***raised in the same species (e.g. goat)***
  - ***added sequentially***
- ***Amplification system:***
  - ***ABC-HRP***
- ***Chromogens:***
  - ***DAB***
  - ***New Fuchsin***

1. Perform any necessary pre-treatments to the tissue sections (dewaxing of tissue sections, antigen retrieval, etc.).
2. Gently wash the slides for 3 × 5 min in TBS containing 0.025% Triton-X 100 (see p 65)
3. Block non-specific protein–protein interactions in TBS containing 10% normal serum, 1% BSA, 0.3 mM glycine and 0.025% Triton-X 100 for 1 h at room temperature (see p 65)
4. Briefly drain the slides (without washing) and apply the first primary antibody optimally diluted in TBS containing 0.025% Triton-X 100 and 1% BSA. Incubate overnight at 4 °C (see p 66)
5. Gently wash the slides for 3 × 5 min in TBS containing 0.025% Triton-X 100
6. Incubate sections with AP conjugated secondary antibody (raised against the species of the first primary antibody), optimally diluted in TBS containing 1% BSA and 0.025% Triton-X 100, for 1 h at room temperature (see p 66)
7. Gently wash the slides for 3 × 5 min in TBS containing 0.025% Triton-X 100
8. Incubate sections with New Fuchsin chromogen for 5–10 min at room temperature (see p 67)
9. Gently wash the slides for 3 × 5 min in TBS containing 0.025% Triton-X 100
10. Apply the second primary antibody optimally diluted in TBS containing 0.025% Triton-X 100 and 1% BSA. Incubate overnight at 4 °C
11. Gently wash the slides for 3 × 5 min in TBS containing 0.025% Triton-X 100

12. Perform endogenous biotin blocking at this point, according to the kit manufacturer's instructions (see p 64)
13. Gently wash the slides for 3 × 5 min in TBS containing 0.025% Triton-X 100
14. Incubate sections with biotinylated secondary antibody (raised against the species of the second primary antibody), optimally diluted in TBS containing 1% BSA and 0.025% Triton-X 100, for 1 h at room temperature (see p 66)
15. Gently wash the slides for 3 × 5 min in TBS containing 0.025% Triton-X 100
16. Quench endogenous peroxidases by incubating the sections with 1.6% H_2O_2 made up in TBS containing 0.025% Triton-X 100, for 30 min at room temperature. While this is incubating, make up the ABC complex (HRP conjugated) according to the manufacturer's instructions (see p 67)
17. Gently wash the slides for 3 × 5 min in TBS containing 0.025% Triton-X 100
18. Incubate sections with the ABC complex. While this is incubating, make up the DAB chromogen according to the manufacturer's instructions (see p 66)
19. Gently wash the slides for 3 × 5 min in TBS containing 0.025% Triton-X 100
20. Incubate sections with DAB for 5–10 min at room temperature
21. Wash for 10 min in running domestic water supply
22. Perform any necessary counterstains (see p 67)
23. Mount in suitable mounting media (see p 67)
24. Analyse using brightfield microscopy

Protocol 4 – Simultaneous enzymatic: unconjugated primary antibodies raised in different species

NB: This protocol uses the same reagents as protocol 3, but has been adapted to demonstrate a simultaneous method.

- ***Primary antibodies:***
  - ***2×, raised against different antigens***
  - ***raised in different species (e.g. mouse and rabbit)***
  - ***added simultaneously***
  - ***unconjugated***
- ***Secondary antibodies:***
  - ***one directly conjugated to AP (raised against the species of the first primary antibody)***
  - ***one directly conjugated to biotin (raised against the species of the second primary antibody)***
  - ***raised in the same species (e.g. goat)***
  - ***added sequentially***
- ***Amplification system:***
  - ***ABC-HRP***
- ***Chromogens:***
  - ***DAB***
  - ***New Fuchsin***

1. Perform any necessary pre-treatments to the tissue sections (dewaxing of tissue sections, antigen retrieval, etc.)
2. Gently wash the slides for 3 × 5 min in TBS containing 0.025% Triton-X 100 (see p 65)
3. Block non-specific protein–protein interactions in TBS containing 10% normal serum, 1% BSA, 0.3 mM glycine and 0.025% Triton-X 100 for 1 h at room temperature (see p 65)
4. Briefly drain the slides (without washing) and apply both primary antibodies, optimally diluted in TBS containing 0.025% Triton-X 100 and 1% BSA. Incubate overnight at 4 °C (see p 66)
5. Gently wash the slides for 3 × 5 min in TBS containing 0.025% Triton-X 100
6. Perform endogenous biotin blocking at this point, according to the kit manufacturer's instructions (see p 64)
7. Gently wash the slides for 3 × 5 min in TBS containing 0.025% Triton-X 100
8. Incubate sections with both secondary antibodies, optimally diluted in TBS containing 1% BSA and 0.025% Triton-X 100, for 1 h at room temperature (see p 66)
9. Gently wash the slides for 3 × 5 min in TBS containing 0.025% Triton-X 100
10. Quench endogenous peroxidases by incubating the sections with 1.6% H_2O_2 made up in TBS containing 0.025% Triton-X 100, for 30 min at room temperature. While this is incubating, make up the ABC complex (HRP conjugated), according to the manufacturer's instructions (see p 67)
11. Gently wash the slides for 3 × 5 min in TBS containing 0.025% Triton-X 100
12. Incubate sections with the ABC complex
13. Gently wash the slides for 3 × 5 min in TBS containing 0.025% Triton-X 100
14. Incubate sections with New Fuchsin chromogen for 5–10 min at room temperature (see p 66). While this is incubating, make up the DAB chromogen, according to the manufacturer's instructions (see p 66)
15. Gently wash the slides for 3 × 5 min in TBS containing 0.025% Triton-X 100
16. Incubate sections with DAB for 5–10 min at room temperature
17. Wash for 10 min in running domestic water supply
18. Perform any necessary counterstains (see p 67)
19. Mount in suitable mounting media (see p 67)
20. Analyse using brightfield microscopy

Protocol 5 – Sequential fluorescence: using conjugated Fab fragments as blocking agents when using unconjugated primary antibodies raised in the same species (see p 113)

- ***Primary antibodies:***
  - ***2×, raised against different antigens***
  - ***raised in the same species (e.g. mouse)***
  - ***added sequentially***
  - ***unconjugated***

- ***Secondary antibodies:***
  - ***one Fab fragment directly conjugated to the first reporter label (e.g. Alexa Fluor® 488)***
  - ***one whole molecule antibody directly conjugated to the second reporter label (e.g. Alexa Fluor® 594)***
  - ***raised against the species of the primary antibodies***
  - ***raised in the same species (e.g. goat)***
  - ***added sequentially***

1. Culture and fix cells optimally (see p 37)
2. Rinse cells for 3 × 5 min in PBS containing 0.1% Tween-20 (see p 65)
3. Permeabilize cells using PBS containing 0.1%Triton X-100 for 10 min at room temperature (see p 65)
4. Block non-specific protein–protein interactions with PBS containing 0.1% Tween-20, 1% BSA and 0.3 mM glycine for 30 min at room temperature (see p 65)
5. Incubate cells with the first optimally diluted unconjugated primary antibody, made up in PBS containing 0.1% Tween-20 and 1% BSA for 1 h at room temperature, or overnight, at 4 °C (see p 66)
6. Incubate cells with the optimally diluted Fab fragment secondary (directly conjugated to the first reporter label)*, made up in PBS containing 0.1% Tween-20 and 1% BSA for 1 h at room temperature, or overnight, at 4 °C (see p 113)
7. Rinse cells for 3 × 5 min in PBS containing 0.1% Tween-20
8. Incubate cells with the second optimally diluted unconjugated primary antibody, made up in PBS containing 0.1% Tween-20 and 1% BSA for 1 h at room temperature, or overnight, at 4 °C (see p 66)
9. Rinse cells for 3 × 5 min in PBS containing 0.1% Tween-20
10. Incubate cells with the optimally diluted whole molecule secondary (directly conjugated to the second reporter label), made up in PBS containing 0.1% Tween-20 and 1% BSA for 1 h at room temperature, or overnight, at 4 °C (see p 66)
11. Rinse cells for 3 × 5 min in PBS containing 0.1% Tween-20
12. Incubate cells with suitable counterstain(s), made up according to the manufacturer's instructions (see p 67)
13. Rinse cells for 3 × 5 min in PBS containing 0.1% Tween-20
14. Mount in a suitable mounting media and analyse by fluorescence microscopy (see p 67)

*A large concentration of labelled Fab fragments may be required to achieve adequate blocking of the first unconjugated primary antibody. However, such high concentrations may induce non-specific binding of the labelled Fab fragments, leading to background staining. If this occurs, reduce the concentration of the labelled Fab fragments, followed by subsequent incubation with unconjugated Fab fragments. The exact concentrations of each will have to be experimentally determined.

Be aware that Fab fragment aggregates bound to the first unconjugated primary antibody may behave as divalent or polyvalent antibodies, with the ability to bind to the whole molecule secondary antibody, leading to unwanted cross-reactivity and therefore false localization of signal.

Protocol 6 – Sequential fluorescence: using unconjugated Fab fragments to effectively change the species in which an unconjugated primary antibody is raised (see p 114)

- ***Primary antibodies:***
  - ***2×, raised against different antigens***
  - ***raised in the same species (e.g. mouse)***
  - ***added sequentially***
  - ***unconjugated***
- ***Secondary antibodies:***
  - ***one Fab fragment unconjugated (raised against the species of the primary antibodies) "A"***
  - ***one whole molecule antibody directly conjugated to the first reporter label (e.g. Alexa Fluor® 488), raised against the species of the Fab fragment (e.g. goat) "B"***
  - ***one whole molecule antibody directly conjugated to the second reporter label (e.g. Alexa Fluor® 594), raised against the species of the primary antibodies and raised in the same species as the other whole molecule antibody directly conjugated to the first reporter label (e.g. rabbit) "C"***
    - ***added sequentially***

1. Culture and fix cells optimally (see p 37)
2. Rinse cells for 3 × 5 min in PBS containing 0.1% Tween-20 (see p 65)
3. Permeabilize cells using PBS containing 0.1%Triton X-100 for 10 min at room temperature (see p 65)
4. Block non-specific protein–protein interactions with PBS containing 0.1% Tween-20, 1% BSA and 0.3 mM glycine for 30 min at room temperature (see p 65)
5. Incubate cells with the first optimally diluted unconjugated primary antibody, made up in PBS containing 0.1% Tween-20 and 1% BSA for 1 h at room temperature, or overnight, at 4 °C (see p 66)
6. Rinse cells for 3 × 5 min in PBS containing 0.1% Tween-20
7. Incubate cells with the optimally diluted unconjugated Fab fragment secondary 'A' (raised against the species of the first primary antibody; e.g. goat anti-mouse H+L), made up in PBS containing 0.1% Tween-20 and 1% BSA for 1 h at room temperature, or overnight, at 4 °C (see p 114)
8. Rinse cells for 3 × 5 min in PBS containing 0.1% Tween-20

9. Incubate cells with the first optimally diluted whole molecule secondary 'B' (directly conjugated to the first reporter label and raised against the species of the Fab fragment; e.g. rabbit anti-goat H+L, Alexa Fluor® 488 conjugated), made up in PBS containing 0.1% Tween-20 and 1% BSA for 1 h at room temperature, or overnight, at 4 °C (see p 66)
10. Rinse cells for 3 × 5 min in PBS containing 0.1% Tween-20
11. Incubate cells with the second optimally diluted unconjugated primary antibody, made up in PBS containing 0.1% Tween-20 and 1% BSA for 1 h at room temperature, or overnight, at 4 °C (see p 66)
12. Rinse cells for 3 × 5 min in PBS containing 0.1% Tween-20
13. Incubate cells with the second optimally diluted whole molecule secondary 'C' (directly conjugated to the second reporter label and raised against the species of the second primary antibody; e.g. rabbit anti-mouse H+L, Alexa Fluor® 594 conjugated), made up in PBS containing 0.1% Tween-20 and 1% BSA for 1 h at room temperature, or overnight, at 4 °C (see p 66)
14. Rinse cells for 3 × 5 min in PBS containing 0.1% Tween-20
15. Incubate cells with suitable counterstain(s), made up according to the manufacturer's instructions (see p 67)
16. Rinse cells for 3 × 5 min in PBS containing 0.1% Tween-20
17. Mount in a suitable mounting media and analyse by fluorescence microscopy (see p 67)

REFERENCES

1. Christensen, N. and Winther, L. (2009) Immunohistochemical staining methods – chapter 15, in *Dako*, 5th edn (eds G.L. Kumar and L. Rudbeck), Dako North America, Carpinteria, California.
2. van der Loos, C. (2009) User protocol: practical guide to multiple staining. *Biotechniques*.
3. Robinson, J.P. and Sturgis, J. (2009) Immunohistochemical staining methods – chapter 10, in *Dako*, 5th edn (eds G.L. Kumar and L. Rudbeck), Dako North America, Carpinteria, California.
4. Atherton, A.J. and Clarke, C. (2001) Multiple labelling techniques for fluorescence microscopy. *Methods in Molecular Medicine*, **57**, 41–8.
5. Manning, C. *et al.* (2012) Benefits and pitfalls of secondary antibodies: why choosing the right secondary is of primary importance. *PLoS ONE*, **7** (6), e38313.
6. Burry, R. (2011) Controls for immunocytochemistry: an update. *Journal of Histochemistry & Cytochemistry*, **59** (1), 6–12.

CHAPTER FIVE

5

Quality Assurance in Immunochemistry

Peter Jackson[1] and Michael Gandy[2]

[1]Department of Histopathology, Leeds General Infirmary, Leeds, UK (retired)
[2]The Doctors Laboratory Ltd, London, UK

INTRODUCTION

Quality control and quality assurance are essential performance measures carried out during the day-to-day operations of diagnostic laboratories. Other measures include ongoing clinical governance, clinical audit, and operational and procedural risk management. All are major requirements necessary to satisfy the accreditation processes required by legislation in many countries.

Quality control within the clinical laboratory (often referred to as internal quality control) is a part of the wider banner of quality assurance. It comprises all of the different measures taken to ensure the reliability of investigations and is not restricted to technical procedures. Quality control is carried out to minimize variability and to ensure the highest quality of laboratory results.

In the laboratory, it can be defined as the analysis of material of known content in order to determine in real time whether the procedures are performing within predetermined specifications. Examples of this are the use of known positive- and negative-control sections, which are stained alongside the test sections, or positive and negative internal

Immunohistochemistry and Immunocytochemistry: Essential Methods, Second Edition. Edited by Simon Renshaw.

controls within the test section, generally non-neoplastic structures which have a known expression levels. If the controls do not demonstrate the expected staining pattern, the test sections are not reported. In this way, quality control controls the release of clinically actionable reports and results.

Quality control differs from quality assessment in that it controls the issue of results in real time on a daily basis. Quality assessment is the challenge of quality control procedures by specimens of known but undisclosed content. It is often done by taking part in external quality assessment schemes (see 'Automated Immunochemical Staining'). It is a retrospective activity that does not affect the day-to-day issue of results. External quality assessment does establish 'between-laboratory' comparability, which in turn allows the identification of best practice and poor performance.

There are several essential elements for quality assurance in immunochemistry.

- Well trained and experienced staff with continuing education and professional development.
- Daily review of all stained slides prepared by the laboratory.
- The use of good quality, well characterized reagents.
- Appropriate antibody titre, antigen retrieval, incubation and detection procedure.
- Appropriate validation for each antibody.
- Well maintained, appropriately calibrated and serviced instrumentation.
- Participation in appropriate external quality assurance programmes and review of returned assessment data.

Standardization of the basic procedures used in histopathology, such as fixation, tissue processing and the treatment of tissue sections prior to staining should make a major contribution in improving laboratory performance. However, tissue processing is far from standardized. The formulation of the fixative used has traditionally been left to each individual laboratory, and similarly tissue processing to paraffin wax is also far from standardized. Laboratories select and employ dehydrating agents of their choice. Tissue processing times vary considerably from laboratory to laboratory with most laboratories having adopted rapid or extended processing times to suit individual tissue types.

This lack of standardization of the fundamental processes of histology culminates in the production of a unique preparation. Subsequently, sections cut from such blocks provide substrates for both tinctorial and immunochemical investigations that may be vastly dissimilar to sections cut from material fixed, processed and sectioned elsewhere. In dealing with such material, it is essential that laboratory personnel understand the problems and pitfalls associated with the basic steps in the preparation of tissue sections for histological investigations and they must be fully aware of the established ways of manipulating protocols in order to give high-quality immunochemical staining on their own and referred sections.

Successful immunochemistry may be seen as the correct integration of several technical parameters. The aim of immunochemistry is to achieve reproducible and consistent demonstration of antigens with the minimum background staining while preserving the integrity of tissue architecture.

METHODS AND APPROACHES

The following are essential considerations for immunochemistry and will each be considered in turn:

- Fixation and tissue processing
- Slide drying
- Microtomy
- Decalcification
- Antigen retrieval techniques
- Immunochemical methodologies
- Controls
- Microscopic interpretation
- Background staining
- Troubleshooting
- Quality assurance schemes

Fixation and Tissue Processing

Adequate and appropriate fixation is the cornerstone of all histological and immunochemical preparations. Tissue fixation should therefore be adequate and appropriate. Standard operating procedures regarding fixation should be rigidly adhered to. Good fixation is the delicate balance between under-fixation and over-fixation. Ideal fixation is the balance between good morphology and good antigenicity. A diagnosis is made on morphological observation and pattern recognition. If the section is of poor quality – and this is usually associated with poor fixation – then diagnosis can be unnecessarily difficult and/or incorrect, leading to inadequate and/or inappropriate treatment.

The demonstration of many antigens depends heavily on the fixative employed and on the immunochemical method selected. Poor fixation, or delay in fixation, causes loss of antigenicity or diffusion of antigens into the surrounding tissue. It is unfortunate that no one fixative is ideal for the demonstration of all antigens, and some tissue antigens necessitate the use of frozen sections. However, the use of frozen material has become less of a requirement with the development and validation of *in vitro* diagnostic (IVD) monoclonal antibodies and staining systems specifically validated for clinical applications on formaldehyde-fixed, paraffin-embedded material.

Cross-Linking Fixatives

In order to achieve the above, a fixative must obtain rapid penetration into the tissue. This is more appropriate when considering formaldehyde, as this is most commonly used on pieces of tissue prior to processing to paraffin wax. Precipitant fixatives tend to be employed on pre-cut tissue sections or cytological preparations.

To ensure rapid penetration of the tissue by any fixative (and hence, rapid fixation), the smallest possible pieces of tissue must be taken from the gross specimen. Although sample size must be taken into consideration, tissue pieces should be no bigger than the tissue cassettes that they are processed in and should fit into the cassettes without being crushed by the lid and allow for free flow of fixative and processing reagent around the sample. Crushing of tissues can cause non-specific immunochemical staining patterns. For gross specimens, tissue penetration can be enhanced further by perfusion or by cutting into the specimen before it is submerged in fixative.

Standardization of fixation in the United States has been addressed by the College of American Pathologists. Their recommendations are that, for optimal results, tissue specimens with maximum dimensions of $1.0 \times 1.0 \times 0.4$ cm should be fixed for a minimum of 3.5 h and a maximum of 18 h in fresh neutral-buffered formaldehyde. The neutral-buffered formaldehyde must be less than 1 month old since formaldehyde is in a gaseous form when dissolved in aqueous solution, so older preparations may have lower concentrations of formaldehyde than fresh. Formaldehyde-fixed tissues that cannot be processed immediately for paraffin embedding should be stored in 70% alcohol until processed, but always bear in mind that alcohol itself is a fixative, so in such cases the conditions should be standardized to further reduce variability in results.

Tissue penetration by formaldehyde is rapid due to its low molecular weight, with initial binding to proteins occurring within 24 h [1]. Although methylene bridge formation can take several days to reach an optimum, with immunochemistry this is not considered so much of a problem. Optimal fixation leads to a high degree of antigen masking, requiring antigen retrieval in order to allow antibody binding. As a result, 24 h is generally considered to be an optimal fixation time, providing an acceptable compromise between obtaining good histological preservation and immunolocalization while keeping antigen masking to a minimum [2]. Tissue intended for processing to paraffin wax should be stored long term in 70% alcohol in order to maintain antigenicity, following the initial 24-h formaldehyde fixation. The same applies to tissues sitting in automated processors over weekends [2].

The further introduction of immunochemical methods to aid in determining prognostic and predictive factors for disease progression and targeted therapies, based on semi-quantitative assessment, has led to a more stringent approach to standardizing the pre-analytical histology process. The stand out examples of this are the ASCO/CAP guideline recommendations for HER2 IHC and ISH testing in breast cancer, whereby samples are recommended to undergo minimal cold ischaemic time followed by fixation in 10% neutral-buffered formaldehyde for a period of 6–72 h depending on sample size and be processed to paraffin using only validated tissue processing methodologies.

Poorly fixed tissue blocks do not process to paraffin wax adequately. Alcohol used in tissue processing dehydration steps is an excellent fixative and an excellent dehydrating agent. However, if asked to perform as both a fixative and a dehydrant, as with poorly fixed tissue blocks, it fails to achieve either and a poorly processed paraffin block ensues. If poor processing occurs regularly within a laboratory, it pays to re-examine the fixation protocols rather than placing the blame solely on the quality of the alcohol. However, in some instances, reprocessing of poorly processed blocks may be carried out. It may be necessary only to re-infiltrate with paraffin wax, to reprocess to xylene or to reprocess to alcohol. In all cases, reasonable immunochemical staining can be achieved for most antigens.

Precipitant Fixatives

Alcoholic fixatives, usually methanol, acetone, or combinations of these, used either alone or in combination with formaldehyde, are often used for whole-cell preparations or for cryostat sections using fresh frozen tissue. However, frozen sections have certain inherent disadvantages when compared with their paraffin counterparts. These include the following:

- Poor morphology in comparison with their paraffin counterparts.
- Poor resolution at higher magnifications.
- Availability of fresh tissue. Special arrangements are required for the collection and storage of fresh unfixed tissues.
- Limited retrospective studies.
- Hazards to health associated with the handling of fresh unfixed material, for example human immunodeficiency virus, hepatitis B virus and tuberculosis.
- Receiving tissue in the laboratory as soon as possible after removal. Good liaison with clinicians and theatre staff is essential.

Alcohol (and acetone) penetrates tissue poorly and is generally only used on tissue sections or cytological preparations as opposed to pieces of tissue. However, some laboratories do utilize alcohol on small pieces of tissue and obtain good cytological definition. Alcohol is often employed in other fixative mixtures to reduce the time spent during tissue processing, for example Carnoy's (methacarn) fixative for demonstrating nucleic acids.

Short acetone fixation alone does not completely stabilize frozen sections against the detrimental effects of long incubations in aqueous solutions. Following long immunochemical staining techniques, acetone-fixed frozen sections may show morphological changes such as chromatolysis and apparent loss of nuclear membranes. It has been observed that such changes to tissue morphology can be prevented by ensuring that the tissue sections are thoroughly dried both before and after fixation in acetone. Treating fresh tissue as described in Protocol 1 will generally produce high-quality immunochemical staining.

Protocol 1 – Preparation of frozen sections for immunochemical staining
Equipment and reagents

- Pre-cooled isopentane in liquid nitrogen
- 3-Aminopropyltriethoxysilane (APES)-coated slides or Superfrost Plus slides (Fisher Scientific)
- Acetone

Method:

1. Snap freeze small pieces of tissue (5 × 3 × 3 mm) using pre-cooled isopentane in liquid nitrogen (see p 36).
2. Cut frozen sections at 5 µm. Pick up on appropriately coated or positively charged slides.
3. Air dry sections overnight at room temperature.
4. Fix sections in acetone (room temperature) for 10 min.
5. Air dry sections.
6. Proceed with immunochemical staining protocol or store sections at −20 °C.

In view of the disadvantaged associates with the production of frozen sections, the general trend is the one away from the use of frozen sections, and most immunochemical investigations performed in diagnostic histopathology laboratories are now carried out on formaldehyde-fixed, paraffin-embedded tissue sections.

Microtomy

It is essential that immunochemical stains are carried out on well-fixed, well-processed, high-quality paraffin sections. Sections should be of nominal thickness (3–4 µm) and free from the artefacts associated with microtomy. Air bubbles under tissue sections can trap antibodies and chromogen (see Fig. 5.1). Knife scores enhance artefacts around the knife tract. Displaced tissue can obscure positively stained cells, and debris from the water bath may make interpretation difficult. Debris can usually be identified from the actual tissue section, as it appears in a slightly different focal plane. If sections are too thick, reagents can be trapped between cell layers, leading to false-positive staining or general background. This also happens in areas where the tissue is folded or creased. Sections for immunochemistry should be picked up on appropriately coated or positively charged slides, otherwise they are unlikely to survive the techniques associated with heat-mediated antigen retrieval.

For immunochemistry, we have seen the move away from the use of direct hotplates to incubate slides due to thermal cycling causing oxidation of certain antigens (e.g. breast hormone receptors). The use of incubators are now readily used to try and preserve (where possible) antigenic structures, which may be sensitive to direct heat or the pulsing nature of thermostatically controlled hotplates. This is particularly important for semi-quantitative immunochemical methods, for example the detection of oestrogen receptors, whereby minimal changes in antigen expression can alter the reported result and potentially change prognosis and therapy.

The two commonly used methods now employed are as follows:

- Adequately drain slides and incubate at 37 °C overnight or
- Adequately drain slides and incubate at 60 °C for 1 h.

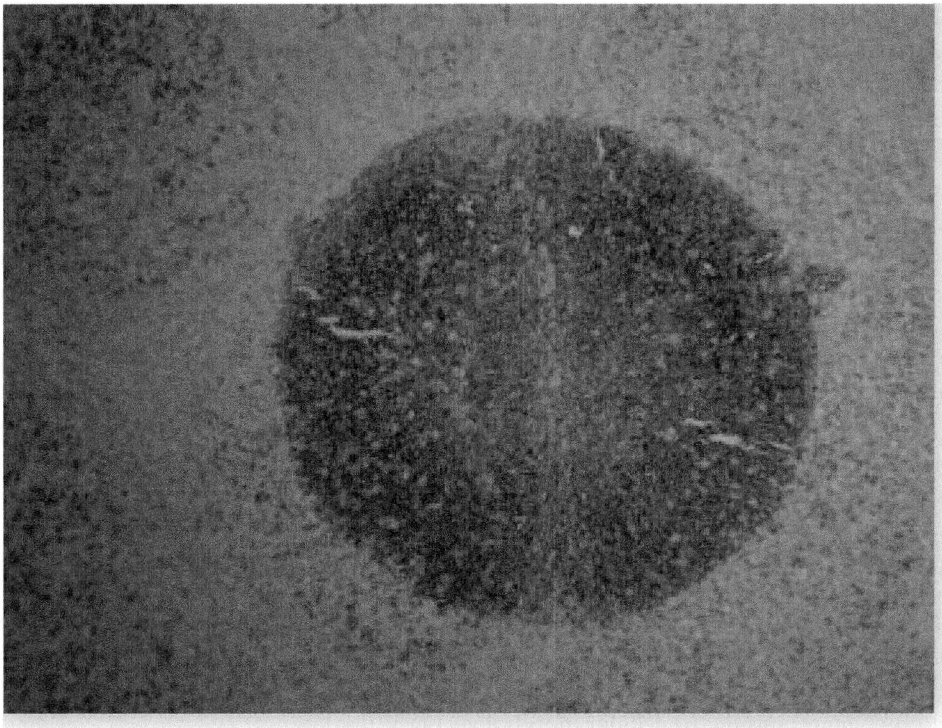

FIGURE 5.1 Example of a Microtomy Artefact: An Air Bubble under the Section Has Trapped Antibodies and Chromogen, Leading to the Observed Non-specific Artefact

For particularly sensitive, low level or labile antigens, slides may be stained following air drying alone. However, these sections may be subject to excessive lifting and other associated artefacts because of poor adherence to the slide.

The duration and intensity of heating tissues during embedding or slide drying should therefore be kept to a minimum and certainly no longer than overnight. As with most histological processes, there is considerable procedural variation between laboratories. It is no exception with slide drying methodologies. For H&E preparations, thermostatically controlled hotplates are readily employed with temperature settings of approximately 5–10 °C above the melting point of the paraffin wax, with slides in direct contact with the heating surface.

Decalcification

It is essential that a specimen is adequately fixed prior to decalcification. Laboratories should choose a decalcifying reagent that has been evaluated in their own laboratory and has been shown not to destroy the antigen in question. This evaluation should be carried out using both tumour and normal tissues. CD15 is an example of a marker that may be destroyed on the tumour by decalcification but may remain detectable on normal cells. However, several researchers have reported that decalcification does not appear to be deleterious to any great degree on the demonstration of numerous antigens. Athanasou [3] observed that decalcification in strong acids did not diminish the reactivity of antigens such as LCA, S100 and EMA. Weaker acids (formic, acetic) and EDTA showed greater preservation of antigenicity with the added advantage of better morphology. TCA showed merit as a one-step fixation/decalcifying agent for both paraffin-embedded and frozen sections. Matthews [4] and Mukai [5] and co-workers demonstrated similar findings on other antigens.

Conversely, Miller [2] claims to have observed damaging effects of decalcification on antigens such as CD43, Ki67, ER and PR and discourages the use of strong acids as decalcifying agents. This further enforces the need for laboratories to choose a decalcifying reagent that has been evaluated in their own laboratory and has been shown not to destroy the antigen in question.

Antigen Retrieval

The process known as antigen retrieval is applied to aldehyde-fixed tissues in which the antigenicity has been reduced by the formation of methylene bridges between components of the amino acid chains of proteins. In many instances, immunoreactivity can be restored without compromising the cellular morphology of the tissues.

On-board antigen retrieval mechanisms are now part and parcel of modern automated immunostainers. However, some laboratories with a small demand for immunocytochemistry may prefer to perform antigen retrieval and subsequent immunostaining by hand.

Proteolytic (Enzymatic) Antigen Retrieval

This can be accomplished through the use of a protease before immunochemical staining. The use of proteolytic enzyme digestion on formaldehyde-fixed, paraffin-embedded tissue sections was first described by Huang et al. [6]. The enzyme-based retrieval method increased the range of useful antibodies that could be applied to routinely fixed, paraffin-embedded tissue sections in diagnostic histopathology. Not all antigens require proteolytic digestion, and care must be taken not to create 'false' antigenic sites or to alter existing antigenic sites. In some instances, immunochemical staining may be reduced or absent following enzyme digestion.

For general use, trypsin is probably the most widely used proteolytic enzyme used for enzymatic antigen retrieval. Care should be taken in the selection of trypsin. Crude and relatively inexpensive trypsin containing the common impurity chymotrypsin often performs best. In fact, chymotrypsin is probably the active ingredient, as purified trypsin gives poor results and chymotrypsin alone can be used successfully [7]. Other popular enzymes include pronase, proteinase K and pepsin. A few antigens such as IgE are said to be preferentially revealed by protease XXIV. The choice of protease, its concentration, and its duration of digestion are largely empirical. It is, however, important to optimize and validate the use of the enzyme of choice rather than to use a broad range of enzymes.

Key factors for enzyme digestion are as follows:

- Temperature
- pH
- Enzyme concentration
- Duration of digestion

For optimum immunochemical staining, digestion time is critical and depends on the duration of fixation in formaldehyde. Tissues fixed in formaldehyde for long periods usually require prolonged exposure to proteolytic enzyme. It may be found that digestion time varies because of batch-to-batch variation in enzyme activity. Another pitfall with enzymatic antigen retrieval is that poorly fixed tissues are easily overdigested with a resulting loss of morphological detail.

Where fixation and paraffin-processing schedules are not known, as with referred blocks and slides, it may be necessary to perform a range of digestion times. In these cases, it is worth recording the optimum digestion time, which may be useful for future immunochemical staining studies on material sent from the same source.

For the above-mentioned reasons, it may not be wise to recommend any one protease over another. Each laboratory must experiment with several proteolytic enzymes using control sections cut from tissue blocks fixed and paraffin processed in their own department in order to identify the optimal conditions for their own material.

In summary:

- Only aldehyde-fixed tissues require protease digestion.
- There are only a limited number of antigens that benefit from the procedure.
- The procedure must be optimized and validated using the laboratory's own material.
- Failure to optimize the protease digestion time to the duration of fixation may produce false-negative results.
- The longer the duration of fixation, the longer the protease digestion time required.

Heat-Induced Epitope (Antigen) Retrieval

Heat-induced epitope (antigen) retrieval (HIER) has proved to be a revolutionary technique, as many antigens previously thought to have been lost or destroyed by fixation and paraffin processing can now be recovered and demonstrated. Antibodies such as MIB1 (Ki67) and oestrogen receptor, which were previously only demonstrable in frozen tissue sections, now work well following heat pre-treatment on paraffin sections.

However, the theory behind HIER remains unclear. To date, there are three hypotheses:

- Heavy metal salts (protein-precipitating fixatives) act as secondary fixatives following primary tissue fixation in formaldehyde and frequently display better preservation of antigens than cross-linking aldehydes.
- During formaldehyde fixation, inter- and intra-molecular cross-linkages alter the protein conformation such that it may not be recognized by some antibodies. HIER removes the weaker Schiff bases but does not remove the methylene bridges. The resulting protein conformation is intermediate between fixed and unfixed.
- Calcium coordination complexes formed during tissue fixation prevent antibodies from combining with epitopes on tissue-bound antigen by steric hindrance. High temperatures weaken or break some of the calcium coordinate bonds, but the effect is reversible on cooling and the complexes re-anneal. The presence of a chelating agent at the particular temperature at which the coordinate bonds are disrupted removes the calcium complexes and hence the steric hindrance.

It is clear that the mechanism(s) of HIER is poorly understood, and as yet there has been no convincing explanation regarding the rationale of the mechanisms involved. However, without the application of such techniques, there is no doubt that immunochemical staining is vastly inferior. This application illustrates the fact that histopathology laboratories often stand alone in the field of medical laboratory science and employ techniques about which little is known to improve staining in tissue sections.

The following are the commonly used methods of HIER:

- Microwave oven heating. This involves the boiling of sections in a suitably microwavable receptacle. Although the microwave oven method allows good reproducible results for many antigens, the limited numbers and the constant attention required to ensure that the sections do not dry out make this method very time consuming. The alternative large-batch microwaving does suffer from inconsistencies, thought to be due to 'hot' and 'cold' spots creating an unbalanced retrieval of antigens, and the rigourous boiling of tissues can lead to dissociation from the slide. Scientific models have features such as stirrers and temperature-monitoring probes to try and alleviate these problems. To help prevent section dissociation from the slide, always use APES-coated or charged slides.
- Pressure cooking. This method does not tend to suffer from such inconsistencies and is far less time consuming. Depending on the capacity of the pressure cooker, up to 75 slides can be retrieved at the same time. Pressure cooker antigen retrieval is carried out under full pressure when the temperature of the retrieval solution reaches 120 °C. The 20 °C difference in temperature between pressure cooker and microwave retrieval appears to be beneficial for the demonstration of many antigens and is the retrieval method of choice in many diagnostic laboratories. The pressure cooker seal should be

changed regularly. A defective seal leads to prolonged boiling. The time taken for the pressure cooker to reach maximum pressure for each individual run should be noted and when times begin to extend the seal should be changed.

- Microwave pressure cooking. This uses a microwaveable pressure cooker, which is brought to full pressure by heating in the microwave oven. Such pressure cookers are relatively inexpensive and can be purchased from leading manufacturers of domestic plastics. Scientific microwave ovens can come equipped with purpose-built, on-board pressurized vessels.
- Automated pressure cooker. A digitally controlled electric pressure cooker. As the heating cycle is controlled digitally, inconsistencies between batches are minimized.
- Steamer. This method involves the use of a domestic vegetable or rice steamer. The antigen retrieval solution is brought to near-boiling temperatures by internally generated steam. The major disadvantage of this method is the extended times of retrieval; however, the use of steam provides a uniform method for heating the antigen retrieval solution.
- Combinations of HIER and proteolytic enzyme digestion. This involves a HIER method followed by a brief digestion in trypsin or protease enzyme. This technique can be used for the so-called difficult antigens and is useful when dealing with bone marrow trephines and renal biopsies to give reliable light-chain demonstration of myeloma and AL amyloid.

There are a wide range of antigen retrieval solutions available, which often makes selection difficult. In any laboratory, it is important that retrieval solutions are evaluated using control sections produced by that particular laboratory. It may be found that three or four retrieval solutions will be selected for use as these may give optimal demonstration for certain antigens of interest. Popular antigen retrieval solutions include citrate buffer (pH6.0) and Tris/EDTA (pH 9.0). They can be divided into two groups:

1. Carboxylic and organic compounds used at pH 6.0–pH 9.0:
   - Citric acid
   - Sodium carbonate
   - Sodium bicarbonate
   - Urea
   - Maleic acid
   - Sodium acetate
   - EDTA
2. Metal and salt solutions:
   - Aluminium chloride
   - Sodium chloride
   - Sodium fluoride
   - Zinc sulfate

- Lead thiocyanate
- Calcium chloride
- Nickel chloride
- Magnesium chloride
- Ammonium chloride

There are a number of advantages and disadvantages of HIER. The main advantage is that heating times to retrieve antigens tend to be standard, regardless of the duration of fixation. This is in contrast to the variability in digestion times required when using proteolytic enzyme digestion. Others advantages include the following:

- Increased intensity of staining and the number of cells stained.
- Demonstration of antigens that are not usually demonstrable in formaldehyde-fixed, paraffin-embedded tissue.
- Production of consistent, reliable, high-quality immunochemical staining of formaldehyde-sensitive antigens.

The main disadvantage when performing HIER is that extreme care must be taken not to allow the sections to dry (so-called 'flash drying'), as this destroys antigenicity and produces section artefacts. Antigen retrieval solutions should be flushed from the container with cold running water. The sections can then be removed when the fluid is cool.

Disadvantages associated with over-retrieval include the following:

- Destruction of antigenicity giving false-negative staining.
- False-positive staining caused by binding other than antibody to antigen.
- The tissue can appear ragged.
- The tissue sections can become dissociated from the slides.
- The tissue may lose its ability to take up counterstain.
- Mucinous areas may be lost.
- Non-specific staining may be increased.

However, the advantages of HIER far outweigh the disadvantages, and therefore it is essential that the techniques involved in HIER are carried out correctly and consistently.

Immunochemical Methodologies

There are numerous immunochemical staining techniques (see p 63). The selection of a suitable technique should be based on parameters such as the type of preparation under investigation, for example frozen section, paraffin or cytological preparation, and the degree of sensitivity required. Many techniques, including the direct method, the indirect method, peroxidase anti-peroxidase and alkaline phosphatase anti-alkaline phosphatase, have now been superseded by more sensitive methods. Good immunochemistry relies on good detection. Many antigens are depleted by the processes involved in fixation and paraffin processing, and it is of the utmost importance to employ a modern, sensitive detection and visualization system.

Major improvements in immunostaining were seen with the introduction of streptavidin- and biotin-based detection systems. Over recent years, the extended polymer-labelled antibody detection systems (often associated with automated immunostainers) such as EnVision (Dako), Bond Polymer Refine (Leica) and UltraView (Ventana) are gaining in popularity. The great advantage of these detection systems is that they are not based on the streptavidin–biotin detection methods, and therefore endogenous biotin does not cause problems in interpretation. Such methods are now regarded as yielding greater sensitivity and a cleaner preparation compared to the traditional ABC methods of the 1990s and have the added advantage that they are quicker and easier to perform. The extended labelled polymer methods are based on either two- or three-layered indirect techniques and utilize pre-diluted and ready-to-use enzyme-labelled secondary or tertiary layer antibodies. These visualization methods are now readily employed on fully automated immunochemical staining systems.

Avidin–Biotin Techniques

These methods utilize the high affinity of the glycoprotein avidin for biotin (vitamin H). Biotin can be conjugated to a variety of biological molecules, including antibodies. One molecule of avidin can combine with four biotin molecules, and this affinity is made use of in the ABC systems. However, avidin has two distinct disadvantages when used in immunochemical detection systems. It has a high isoelectric point of approximately 10 and is therefore positively charged at neutral pH when used in immunochemical staining methods. Consequently, it may bind non-specifically to negatively charged structures, such as the nucleus. The second disadvantage is that avidin is a glycoprotein and reacts with molecules such as lectins via the carbohydrate moiety. This reaction may be blocked by using an analogue of the carbohydrate on the avidin, for example, 0.1 M α-methyl-D-mannoside can be added to the solution containing the avidin.

Both problems may be overcome with the substitution of streptavidin for avidin. Streptavidin is a protein isolated from the bacterium *Streptomyces avidini* and, like avidin, has four high-affinity binding sites for biotin. Streptavidin has an isoelectric point close to neutral pH and therefore will not bind to positively charged structures at the near-neutral pH used in immunochemical staining systems. Furthermore, streptavidin is not a glycoprotein and therefore does not bind to lectins. The physical properties of streptavidin make this protein much more desirable for use in immunochemical staining systems than avidin.

The more sophisticated and sensitive extensions of the streptavidin–biotin technique employ pre-formed complexes. Streptavidin and biotinylated enzyme are simply mixed together at appropriate concentrations and allowed to stand for 30 min to allow the complex to form. Failure to leave the complex for the required amount of time will lead to a weak signal. The pre-formed complex is then added to the biotinylated antibody.

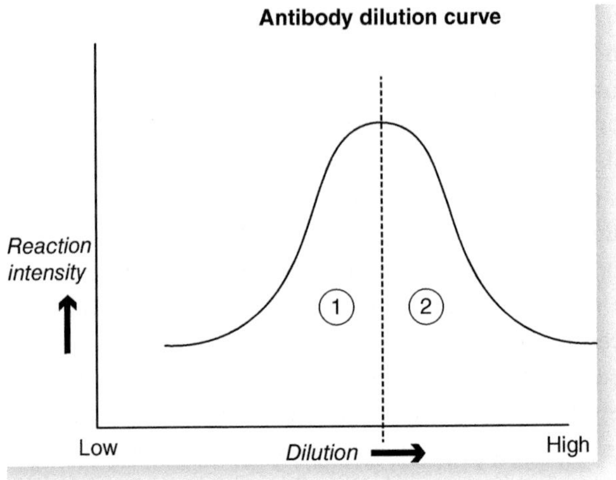

FIGURE 5.2 Determination of Optimal Dilution of Antibody: The Poor Reaction in Area 1 Is due to Steric Hindrance of the Primary Antibody Accessing the Antigen as the Antibody Concentration Being Too High. This Is Known as Prozone, but Is a Rare Phenomenon in Immunochemical Staining. The Suboptimal Reaction in Area 2 Is Caused by Insufficient Primary Antibody, That Is the Concentration of Primary Antibody Is Too Low

Careful stoichiometric control ensures that some binding sites remain free to bind with biotinylated antibody. This allows the pre-formed complex to bind and provides a very high signal at the antigen-binding site. Horseradish peroxidase or alkaline phosphatase can be used as the enzyme label.

As a word of warning, because pre-diluted secondary and tertiary antibodies are titered to work optimally in a specific kit, do not attempt to mix reagents from different kits.

Optimal Dilution of Primary Antibody

The optimum dilution (titre) of primary antibody for immunochemistry is the concentration of the primary antibody that gives the optimal specific staining with the least amount of background staining.

The optimal dilution will depend on the type and duration of fixation. Serial dilutions of antibody will often give the distribution of reactivity as shown in Figure 5.2. The optimum concentration of primary antibody is that measured below the apex of the peak, and the use of several sections with varying densities of antigen will aid the determination of a correct working dilution.

Controls

The importance of using appropriate controls cannot be overstressed! Controls validate immunochemical staining results and can quickly identify problems with reagents, test specimens or an individual's immunochemical staining technique. The results of immunochemical staining are essentially meaningless unless the appropriate controls have been employed.

In any given immunochemical staining run, most laboratories routinely employ a positive tissue control and run alongside an identical section used as a reagent control.

Positive and Negative Tissue Controls

It must be stressed that control tissues must be subjected to exactly the same pre-treatments and immunochemical staining protocol steps as the test tissues in order to get an accurate result.

The inclusion of a positive-control section known to contain the antigen in question is essential every time immunochemical staining is carried out. Without such a control, a negative result on the test material is meaningless because there is no guarantee that the reagents or kit are in good working order and have been applied in the correct order. With the introduction of fully automated immunochemical systems and individual slide positions, it is a requirement that (where possible) same-slide positive controls are used, given that each individual staining position could be subject to its own individual error(s).

In contrast, negative tissue controls do not contain the antigen in question. Any staining must therefore be from the primary antibody (either the wrong one has been used or from specific/non-specific reactions) or non-specific reactions intrinsic to the labelling system.

Although very rarely obtainable, knockout tissues make excellent negative controls. Mice, for example, can be genetically engineered to lack the protein (antigen) under investigation. A negative staining result from a knockout tissue combined with a positive result from a positive-control (wild-type) tissue is a strong indication of specificity. However, both control types have to be from the same species for validity.

Reagent Controls

The primary goal of reagent controls is to ensure that the primary and secondary antibodies recognize their target antigens. All other factors should be kept constant, for example the buffers used to dilute the antibodies, incubation time and temperature.

Where possible, purchasing immunogen-affinity-purified primary antibodies and pre-absorbed secondary antibodies should greatly increase reagent specificity. Primary antibodies should first have their optimal dilutions determined, followed by testing and validation on a panel of tissues known to be both positive and negative for the antigen in question. Reference laboratories who have already successfully demonstrated the antigen(s) in question may be used to evaluate newly introduced markers.

An essential control when evaluating a new polyclonal antibody for demonstrating an antigen in an unknown location is to pre-absorb the primary antibody with an excess of its specific antigen, so that no antibody-binding sites are available to react with the tissue and staining does not occur (see Fig. 5.3). This may be regarded as the ultimate test

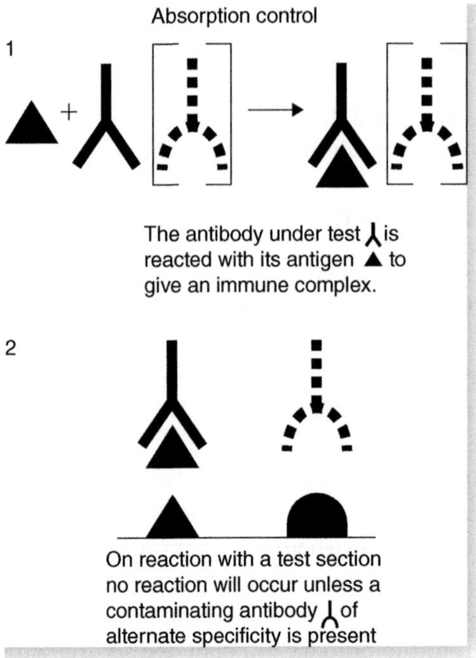

FIGURE 5.3 Example of Absorption Control for Immunochemical Staining

for specificity. If staining does occur, then the staining must be due to a contaminating antibody or non-specific interactions and not to the antigen–antibody interaction under investigation. For reasons of economy, absorption should be carried out at the highest possible dilution of the antibody compatible with consistent unequivocal staining, since the higher the concentration of antibody, the more antigen will be needed for neutralization.

However, if no antigen is available, or if financial or time constraints do not allow, then most laboratories replace the primary antibody under investigation with an isotype control or normal serum from the same species used to raise the primary antibody. An isotype control is defined as an antibody of the same isotype and raised in the same species as the primary antibody under investigation, but to an irrelevant protein, such as keyhole limpet haemocyanin. Any staining in the reagent control must be due to non-specific reactions between the immunoglobulin and the tissue, or from reactions intrinsic to the labelling system. It is important when using an immunoglobulin fraction as a reagent control to always use it at the same immunoglobulin/overall protein concentration as that of the primary antibody. Similarly, both should be of the same age and obtained using the same purification methods. This is because soluble immunoglobulin aggregates present within the serum can exacerbate non-specific staining, and the amount present is influenced by these factors.

However, if the primary antibody is monoclonal, then the only real option is to use an isotype control.

Internal Controls

This type of control is a 'built-in' control and is the best possible control because the variables of tissue fixation, processing and section treatment prior to staining are eliminated. Internal controls contain the target marker in the surrounding normal tissue elements of the test sample. One example is the presence of S100 protein in melanoma and normal tissue elements such as peripheral nerves and melanocytes.

Control Blocks

Control blocks should be selected from a case that expresses (is positive for) the antigen to be immunochemically stained. A negative reagent control section should be cut from the same block.

Care should be taken when using stored or archived tissue sections as positive controls. Several publications have highlighted the dangers in the use of old tissue sections as positive controls. Some antigens in paraffin sections degrade during storage, probably due to oxidation. It is a good practice to cut only enough sections from the control block to suffice for the week's work.

It is important that a section from the control block is stained with haematoxylin and eosin whenever a batch of control slides is cut to ensure that the section contains the lesion of interest. This is particularly important when utilizing tissue microarray (TMA) methodologies, where the small sample (core) size increases the incidence of the selected region not being representative of the donor block. TMAs are a useful way to enable well-characterized control material to last longer by creating multiple control blocks from a single case. The use of small TMA cores also allows control blocks of multiple tissue and tumour types to be made and further facilitates the ability to include both positive and negative antigen control tissue on the same slide as the test tissue.

Microscopic Interpretation

Evaluation and interpretation of the finished product can be the most difficult aspect of immunochemical staining. Assessment of section quality regarding fixation, tissue processing, microtomy and possible artefacts are all taken into consideration together with the demonstration and localization of the antigen.

Whoever interprets the immunochemical staining must know what the stain should look like in both normal tissue and tumour. Assessment of the pattern of staining is important. There are three basic patterns of specific staining that can be observed, namely cytoplasmic, nuclear and surface. The amount of precipitate generated by the labelling enzyme/chromogen combination, and therefore the intensity of the reaction, is proportional to the amount of antigen present. Not all cells contain the same antigen in question. For those that do, there may be varying quantities of the antigen from cell to cell, and they will therefore stain with varying intensity. Staining may also be focal or diffuse. Focal staining is localized in discrete areas of the cell, whereas diffuse staining may occupy larger areas of the cell and also adjacent cells.

Immunohistochemical non-specific background staining is usually associated with collagen and connective tissues, and although non-specific background staining can be minimized by good staining technique, it is sometimes difficult to abolish, particularly when using polyclonal antibodies. More problems in interpretation may arise due to endogenous enzyme activity in red blood cells and macrophages due to poor blocking techniques.

Interpretation is a comparison of the specific and non-specific staining patterns of the test section with that of a control. Evaluation of a stained section should only be carried out when the tissue type, the basic histological procedures, the pre-treatment and the immunochemical staining technique have all been taken into account.

Under-fixation may preserve the antigen being investigated, but the morphology is likely to be poor and may cause difficulties in interpretation. Longer fixation will ultimately give better tissue morphology, but antigens may become altered, masked, depleted or destroyed. Interpretation of staining patterns should be made based on adequately fixed areas of the section. Cells that have undergone autolysis, necrosis or have been crushed may exhibit non-specific staining. Only the staining pattern of viable cells should be considered for interpretation.

Background Staining

The major causes of background staining in immunochemistry are hydrophobic and ionic interactions and endogenous enzyme activity.

Hydrophobic Interactions

Tissue hydrophobicity is increased by formaldehyde (and glutaraldehyde) fixation, leading to an increase in the incidence of general background staining. All proteins are hydrophobic to some degree, with the side-chain amino acids phenylalanine, tryptophan and tyrosine linking together in order to eradicate water. Aldehyde fixation increases hydrophobicity by linking reactive alpha and epsilon amino acids between adjacent proteins or within a single tertiary protein structure. Connective tissues, squamous epithelium and adipocytes (if not adequately removed during tissue processing) are especially prone. Such an increase in hydrophobicity is correlated to fixation temperature, duration and pH, further enforcing the need for an optimized and standardized fixation regime.

Non-specific background staining is most commonly produced because of the attachment of the primary antibody drawn non-immunologically to highly charged groups present on connective tissue elements. Positive staining is not due to localization of the antigen but due to non-specific attachment of the primary antibody to connective tissues. Because the primary antibody is attached to connective tissue moieties, the subsequent labelling (secondary) antibodies will be attracted not only to primary antibodies located on the specific antigen but also to the antibody bound to connective tissues. The most effective way of minimizing non-specific staining is to add an innocuous protein solution, such as bovine serum albumin, to the section before applying the primary antibody. Other techniques are the addition of a detergent, for example Tween-20, to the washing buffer or using a diluent with a low ionic strength. The protein should saturate and neutralize the charged sites, thus enabling the primary antibody to bind only to its intended site. The addition of 0.3 M glycine to blocking buffers utilized on aldehyde-fixed specimens will also help to reduce non-specific binding by capping off free aldehyde groups, rendering them unable to bind to antibody lysine residues.

Traditionally, non-immune serum from the same animal species that produced the secondary antibody is used as a blocking serum, as this also has the effect of binding to any endogenous immunoglobulin in the tissue that shows reactivity with the secondary antibody. In practice, any animal serum or protein can be used for this purpose as long as the protein used as a block cannot be recognized by any of the subsequent antibodies used in the technique.

Ionic Interactions

These occur when proteins of opposite net charges meet. Most IgG class antibodies have isoelectric points ranging from pH 5.8 to 7.3. Most will have a net negative surface charge when presented in the buffers usually used in immunochemical staining, that is in the pH range of 7.0–7.8. Interactions can therefore be expected if tissue proteins have a net positive surface charge. In general, ionic interactions can be reduced by the use of diluent buffers with higher ionic strength or by the addition of 0.1–0.5 M NaCl to the buffer usually used.

It should be noted that hydrophobic binding is reduced by the use of low ionic strength buffer. Ionic binding is reduced by the use of high ionic strength buffer. It can be seen that remedies to reduce one may aggravate the other!

Fc Receptors

In frozen sections and cytological preparations, tissue receptors for the Fc portion of antibodies may give rise to additional problems. These Fc receptors present on several cell types such as macrophages and monocytes as part of the natural immune defence mechanism are largely destroyed in paraffin-embedded tissues by formaldehyde fixation and tissue processing. If necessary, Fab or F(ab)2 fragments of antibodies, which lack the Fc portion, should be used.

Endogenous Peroxidase Activity

Certain cells (in particular red blood cells) naturally contain this oxidoreductase enzyme. By applying hydrogen peroxide to the tissue prior to the staining procedure, endogenous peroxidases can be quenched and reversibly prevented from reacting with the chromogen applied as the last detection step (see p 61). This is due to the initial complex formed between the excess of peroxidase and hydrogen peroxide being catalytically inactive if no electron donor is present. Other blockers include ethanol containing 0.2% hydrochloric acid and methanol containing 1% sodium nitroferricyanide and 0.2% acetic acid. An alternative to blocking endogenous peroxidases is simply to use an alkaline phosphatase label instead of horseradish peroxidase.

Thick tissue sections (60 μm) in a free-floating format (i.e. not adhered to a slide; e.g. gelatin embedded) can be treated with a different method using sodium borohydrate, which penetrates the thick tissue sections more efficiently than hydrogen peroxide.

Protocol 2 – Blocking endogenous peroxidases in free-floating, thick tissue sections (60 mm)

Equipment and reagents

- 1% (w/v) Sodium borohydrate solution
- Extraction cabinet
- Magnetic stirrer

Method

1. Wash the free-floating, thick tissue sections in PBS or TBS to remove excess fixative. Carry out steps 2–6 in an extraction cabinet (no humidity).[a]
2. Prepare a fresh solution of 1% (w/v) sodium borohydrate in water: weigh 1 g of sodium borohydrate and add it to a suitable container containing 100 mL of double-distilled water in one movement. Close the lid immediately and stir gently using a magnetic stirrer.
3. Wait 5 min for the sodium borohydrate to dissolve in the water. Effervescence should be observed and the reaction will be exothermic.
4. Open the container and add the solution to the reaction vessels containing the sections.
5. Incubate the sections with this solution for up to 10 min. Effervescence will be observed from the tissue due to oxygen release.
6. Wash the sections 10 times in PBS or TBS until no effervescence is seen.
7. Continue with the immunochemical staining protocol.

Notes

[a]Sodium borohydrate is extremely dangerous as it is an explosive. It must be handled in a dry environment prior to being added to the water, as contact with water releases highly flammable gases, hence, the strict use of an extraction cabinet. Dispose of the used sodium borohydrate solution according to material safety datasheet (MSDS) guidelines.

Protocol 3 –

For the blocking endogenous peroxidases in 4–10 μm thick tissue sections or cytological preparations, please see p 65.

Endogenous Phosphatase Activity

This enzyme is found endogenously in liver, intestine, bone, placenta, leucocytes and certain carcinomas. Applying levamisole at a concentration of 1–2 mM in the chromogen solution will block endogenous phosphatase in all tissues except the intestine, which exhibits a different isoform [8]. Alternative endogenous alkaline phosphatase-blocking solutions include Bouin's fixative and 20% acetic acid. An alternative to blocking intestinal phosphatase is simply to use a horseradish peroxidase label instead of alkaline phosphatase.

Protocol 4 – Blocking endogenous phosphatases in 4–10 µm tissue sections or cytological preparations

Reagents

- Alkaline phosphatase chromogen solution containing 1 mM levamisole (Sigma, catalogue no. L9756)

Method

1. Use a standard immunochemical staining protocol, such as that given in p 68.
2. Apply the alkaline phosphatase chromogen solution containing 0.24 mg/mL (1 mM) levamisole.
3. Incubate for as long as necessary to see specific staining.

Protocol 5 – Blocking endogenous intestinal phosphatases in 4–10 µm tissue sections or cytological preparations

Reagents

- Absolute methanol containing 20% acetic acid (v/v).

Method

1. Use a standard immunochemical staining protocol, such as that given in p 68.
2. Apply absolute methanol containing 20% acetic acid (v/v) for 5 min (see step 10 in the standard immunochemical staining protocol quoted above) instead of the H_2O_2.
3. Continue with the immunochemical staining protocol.

Biotin

Liver, kidney and lymphoid tissues contain high levels of this vitamin B_7. It can also be found in adipose tissue and central nervous tissues in low amounts.

Biotin is often used in immunochemical staining methods to amplify the signal (see p 64). It is therefore important to block endogenous biotin when an ABC amplification system is being used, since the avidin, when applied to the tissue or cells during the amplification step, can potentially bind to the endogenous biotin. To block endogenous biotin, it is recommended to pre-incubate the tissue sections in 0.01% (w/v) avidin (to bind to the endogenous biotin) followed by 0.001% (w/v) biotin (to block any vacant biotin-binding sites on the avidin).

Protocol 6 – Blocking endogenous biotin in 4–10 μm tissue sections

The following steps should be performed before incubation with the biotinylated antibody.

Reagents

- PBS or TBS containing 0.01% (w/v) avidin (Sigma, catalogue no. A9275)
- PBS or TBS containing 0.001% (w/v) biotin (Sigma, catalogue no. B4501)
- PBS or TBS

Method

1. Prepare fresh solutions of PBS or TBS containing 0.01% (w/v) avidin, and PBS or TBS containing 0.001% (w/v) biotin.
2. Apply the PBS or TBS containing 0.01% (w/v) avidin to the tissue sections for 10 min.
3. Rinse the slides in PBS or TBS for 3 × 5 min.
4. Apply PBS or TBS containing 0.001% (w/v) biotin to the tissue sections for 10 min.
5. Rinse the slides in PBS or TBS for 3 × 5 min.
6. Continue with the immunochemical staining protocol.

A very effective biotin-blocking solution for heavily biotinylated sites (which can be very difficult to block using the standard blocking solutions) can also be made in-house [9].

Protocol 7 – Blocking heavily biotinylated sites in 4–10 μm tissue sections

The following steps should be performed before incubation with the biotinylated antibody.

Reagents

- 200 mL distilled water containing two egg whites (solution A)
- Distilled water containing 0.2% (w/v) biotin (Sigma, catalogue no. B4501) (solution B)
- PBS or TBS

Method

1. Apply solution A to the tissue sections for 20 min.
2. Rinse the slides in TBS for 3 × 5 min.
3. Apply solution B to the tissue sections for 20 min.
4. Rinse the slides in TBS for 3 × 5 min.
5. Continue with the immunochemical staining protocol.

Auto-Fluorescence

Auto-fluorescence is the ability of a tissue or cellular component to fluoresce naturally. This occurs independent of the binding of immunochemical staining reagents conjugated to fluorochromes. Auto-fluorescence can lead potentially to false-positive staining. Three types of auto-fluorescence are common:

- Elastin and collagen
- Lipofuscin
- Aldehyde (fixative) induced

Elastin and Collagen Auto-Fluorescence— Elastin and collagen are common components of blood vessels. Among other fluorophores, elastin contains a naturally occurring cross-linking tricarboxylic amino acid with a pyridinium ring [10]. Collagen possesses a similar fluorophore.

Cowen et al. [11] used a dye called pontamine sky blue to successfully quench auto-fluorescence in carotid arteries and mesenteric vessels using a fluorescence resonance energy transfer method. This method only works when using a fluorochrome that emits light in the green spectra (FITC, Alexa Fluor® 488, etc), as the pontamine sky blue shifts the fluorescent emission of elastin and collagen from green to red.

Protocol 8 – Quenching auto-fluorescence due to elastin and collagen in tissue sections [11]

Reagents

- PBS or TBS containing 0.5% (w/v) pontamine sky blue

Method

1. Use a standard immunochemical staining protocol, such as that given in p 68.
2. Apply PBS or TBS containing 0.5% (w/v) pontamine sky blue for 10 min at the step before incubation with the reagent conjugated to the FITC fluorochrome.
3. Rinse the slides in PBS or TBS for 5 min.
4. Continue with the immunochemical staining protocol.

Lipofuscin Auto-Fluorescence— Naturally fluorescent lipofuscin pigment accumulates with age in the cytoplasm of numerous cell types, including those of the central nervous system [12, 13]. Commonly referred to as ageing pigment, it is yellow brown in appearance and accumulates within lysosomes due to peroxidation of lipids [6]. It has an excitation and emission spectrum similar to that of FITC.

Protocol 9 – Quenching auto-fluorescence due to lipofuscin in tissue sections [12, 13]

Reagents

- 70% (v/v) ethanol containing 1% (w/v) sudan black B (stirred in the dark for 2 h)

Method

1. Use a standard immunochemical staining protocol, such as that given in p 68.
2. Apply 70% (v/v) ethanol containing 1% (w/v) sudan black B for 10 min at the step before incubation with the reagent conjugated to the fluorochrome.
3. Quickly rinse the slides 10 times in TBS or PBS.
4. Continue with the immunochemical staining protocol.

Aldehyde-Induced Auto-Fluorescence— Auto-fluorescent compounds are formed between aldehyde fixatives and proteins or amines, with glutaraldehyde fixation giving a greater degree of auto-fluorescence than formaldehyde due to extensive cross-linking. Aldehyde-induced auto-fluorescence is more diffuse and more generalized than specific staining, its intensity increasing with incubation temperature and duration (see p 37 for further information on fixation).

An effective means of avoiding aldehyde-induced auto-fluorescence is simply to avoid using an aldehyde fixative, although this can have certain disadvantages (see p 41 for a discussion of alternative fixatives).

Protocol 10 – Quenching auto-fluorescence in aldehyde-fixed tissue sections or cytological preparations [14]

Equipment and reagents

- 0.1% (w/v) sodium borohydrate solution
- Extraction cabinet
- Magnetic stirrer

Method

1. Use a standard immunochemical staining protocol, such as that given in p 68.
   Carry out steps 2–5 in an extraction cabinet (no humidity).[a]
2. Prepare a fresh solution of 0.1% (w/v) sodium borohydrate in water: weigh 100 mg of sodium borohydrate and add it to a suitable container containing 100 mL of double-distilled water in one movement. Close the lid immediately and gently stir using a magnetic stirrer.
3. Wait 5 min for the sodium borohydrate to dissolve in the water. Effervescence should be observed and the reaction will be exothermic.
4. Open the container and apply this solution to the tissue section or cells (while still effervescing)[b]. For cell monolayers, incubate twice for 4 min each using fresh solution each time. For paraffin-embedded sections, incubate three times for 10 min each using fresh solution each time. Effervescence will be observed from the tissue due to oxygen release.

5. Rinse for 3 × 5 min in PBS or TBS to remove all traces of sodium borohydrate.
6. Continue with the immunochemical staining protocol.

Notes
[a]Sodium borohydrate is extremely dangerous as it is an explosive. It must be handled in a dry environment prior to being added to the PBS or TBS, as contact with water releases highly flammable gases, hence, the strict use of an extraction cabinet. Dispose of the used sodium borohydrate solution according to MSDS guidelines.
[b]Sodium borohydrate reduces aldehyde-induced auto-fluorescence by reducing Schiff bases ($R^1HC{=}NR^2$) that are formed from the reaction between the aldehyde and NH_2 groups.

Formaldehyde (Formalin) Pigment— Specimens fixed in acidic formaldehyde can demonstrate formaldehyde pigment, a brown-black deposit created when acidic formaldehyde reacts with blood. This pigment is related to the acid haematins and can be mistaken for immunochemical staining when diaminobenzidine (DAB) is used as the chromogen (substrate) (see p 46). Formation of formaldehyde pigment can be almost eliminated by fixing in neutral-buffered formaldehyde (pH 7.0) [15].

AUTOMATED IMMUNOCHEMICAL STAINING

The introduction of complete automation in immunochemical staining has provided the opportunity to further standardize immunochemical techniques. The majority of diagnostic testing centres will now utilize a form of staining automation to deliver their routine diagnostic service. Automation includes de-paraffinization, antigen retrieval, immunochemical labelling, antigen visualization and counterstaining, producing a completed preparation requiring only dehydration, clearing and mounting. The main advantages of these systems are as follows:

- Time saving: end users can speed there time doing other things
- Rapid turn-around of immunochemical straining runs, with the possibility of overnight runs
- Pre-made reagents that have passed stringent QC checks
- Improved reproducibility of results due to elimination of sources of human error and variation

The main disadvantages are as follows:

- Expensive to buy (both the systems and the reagents)
- Most systems are relatively inflexible with regard to the immunostaining protocols that can be performed upon them and the reagents that can be used (end users can often only use reagents provided by the system manufacturer)

Improvements in reagent manufacture by industry has enabled the introduction of a wide range of 'ready-to-use' reagents ranging from bulk solutions of dewaxing, antigen retrieval and wash buffer solutions, as well as primary antibodies and detection systems. Automated reagent handling has enabled consistent volume dispensing and uniform tissue coverage, leading to improved staining reproducibility in immunochemical preparations.

Although the order of reagent dispensing remains the same, reagent kinetics and enclosed staining and gradual improvements in reagent manufacture and labelling sensitivity have led to the shortening of staining protocols

The use of vendor-specific pre-programmed steps that include the intermediate washing steps provides ideal starting points for most types of staining protocols, including alkaline phosphatase and automated serial and parallel and double labelling. However, pre-programmed protocols should only be used following appropriate internal validation ensuring diagnostic accuracy of the method adopted with each clinical testing centre.

Artefacts of Fully Automated Staining

Gradient Staining

This artefact occurs with staining systems that operate a linear flow of reagent across the tissue in an enclosed environment and most often occurs in conjunction with high expressing tissue antigens. Primary antibody and subsequent labelling systems react rapidly with high levels of tissue antigen present at the leading edge of reagent flow producing strong initial stain. Subsequent tissue beneath the leading edge often appears weaker in contrast to the upper edge, which is often strongly stained, giving an effect of gradient staining across the tissue section.

Tissue Damage

This artefact can occur with staining systems that mix reagents on top of the tissue section on the instrument. Mechanical damaging due to vortexing may require additional tissue adherence to slides. Coated or charged slides are strongly recommended for use with these systems.

Reagent Precipitate

This artefact occurs due to trapping of reagent under the edges of reagent chambers, each subsequent label then builds up to cause an often linear, brown-coloured staining artefact on the stain upon the stained slide. This debris can then be washed over the section making interpretation difficult.

Inadequate Reagent Flow

This artefact may be caused by obstruction within the reagent chamber (e.g. disaggregated tissue) or by use of non-proprietary or incorrectly formulated reagents. Most automated systems will employ surfactant within wash buffers to facilitate even spreading of reagents. This artefact can also be caused due to the hydrophobic nature of the surface of certain glass slides or by faults in staining platforms that utilize a 'capillary gap' system.

TROUBLESHOOTING

Troubleshooting immunochemical staining procedures is not an easy task. With so many variables, it is often difficult to know exactly where to begin. The best approach is to start with the simple and progress to the more technically demanding issues later:

- Start by ensuring that none of the steps of the immunochemical staining protocol have been omitted, according to whatever detection system is being employed.
- Ensure that all relevant antigen retrieval/blocking steps have been performed.
- Consult the datasheet for the primary antibody to verify that it will indeed work in the specimen type that is being used, for instance, frozen sections, paraffin-embedded sections or cytological preparations. The datasheet for the primary antibody may also give extremely helpful information regarding optimal specimen fixation conditions and antigen retrieval.
- Check that all of the reagents are compatible with each other, for instance, ensure that the secondary antibody detects the immunoglobulin species and subclass of the primary antibody.
- If no apparent problems can be found with the above, check that all of the reagents used have been made up correctly, for instance, that the buffer constituents and pH are correct.
- Check that none of the reagents has gone out of date or has been stored incorrectly.

Table 5.1 is designed to give the user insight into the possible causes of error during an immunochemical staining procedure and the possible solutions, grouped by the observed staining pattern seen in the test specimen and the positive and negative tissue/reagent controls. It is by no means exhaustive!

If no solution can be found after consulting this table, it is often worthwhile contacting the technical department of the primary antibody supplier to see whether they can be of assistance. Failing this, the user may have the luxury of being able to hand over their primary antibody to a fellow colleague to see what results they can obtain. This is especially useful when they carry out the work in a different laboratory, where the laboratory's own protocols, reagents and controls are used. They should immunochemically stain the user's own test specimen and controls alongside their own, subjecting them to the same immunochemical staining protocol and pre-treatment. If they manage to obtain the desired staining pattern, then comparing and contrasting the techniques, reagents, test specimens and controls used should provide (or move the user very close to) the answer.

TABLE 5.1 Troubleshooting Immunochemical Staining Procedures

Weak or absent immunochemical staining in both control and test specimens, with minimal or very little background staining	
Possible Cause	***Possible Remedy***
Paraffin section (cytological preparation if PEG protected) is not dewaxed properly	Ensure an adequate time in the dewaxing solution
Tissues exposed to too high a temperature during embedding or paraffin section drying	The wax bath and oven temperature should not exceed 60 °C (see p 128)
Fixative used is incompatible with the antigen	Refer to the antibody datasheet for possible fixative recommendations
Reagent(s) omitted or applied in wrong order	Follow the immunochemical staining protocol carefully
Incompatible reagents	Ensure that the primary antibodies detect antigen in the species and sample type (frozen, formaldehyde fixed, paraffin embedded, etc.) being studied
	Ensure that the secondary antibodies detect the primary antibody subclass and isotype
	If using an ABC system, ensure that the secondary antibody is biotinylated
	Ensure that the chromogen is compatible with the immunoenzyme label
Reagent(s) made up incorrectly	Follow the manufacturer's instructions and protocol recipes carefully
Reagent(s) too old or degraded	Observe the manufacturer's expiration dates and storage conditions
	Ensure that frozen antibodies are not subjected to excessive freeze–thaw cycles
	Ensure that conjugated antibodies are not frozen
Reagent(s) used at too low/high a concentration	Perform dilution ranges for the antibodies used at each stage to find the optimum (see p 60). Note: at too high a concentration, antibodies may exhibit 'prozone' effects (see p 136)
	Observe the manufacturer's recommended working dilutions/concentrations for commercial reagents
Reagent(s) not incubated for long enough	Increase the incubation times of reagents, in particular for the primary antibody
Reagent(s) not incubated at a high enough temperature	Increase the incubation temperature by a few degrees
	Allow all reagents to warm to room temperature before starting the incubation

TABLE 5.1 (*Continued*)

Antigen not present at detectable levels	Use a more sensitive detection system (see p 63)
	Increase the incubation times of reagents, in particular for the primary antibody
	Antigen may be masked by formaldehyde fixation, so perform antigen retrieval (see p 50) if not already done
	If antigen retrieval has already been performed, check that the reagents have been made up correctly, increase retrieval times, increase heat during HIER, check for defective or underactive performance of the enzyme during enzymatic retrieval, or try another retrieval technique
	Use a different (often non-cross-linking) fixative
Wrong buffer/buffer constituents/pH used.	Avoid PBS when using phosphatase labels
	Avoid sodium azide when using peroxidase labels. Note: diluent pH can affect antibody affinity for the antigen
	Detergent in buffers can remove some membrane-bound antigens
Dissociation of reagents during buffer washes	Use antibodies at lower dilutions
	Replace primary antibody with antibody of the same specification but with a higher affinity for the antigen
Enzyme or chromogen precipitate is soluble in nuclear counterstain, dehydrating solution, or mounting media	Use a non-alcohol-containing counterstain such as Mayer's
	Use a non-alcohol-soluble chromogen such as DAB
	Use aqueous mounting medium
Photo-bleaching occurs when using fluorescent labels	Use a second-generation fluorophore, decrease light/laser intensity during microscopy
Using a mountant containing glycerol when using phycobiliprotein fluorescent labels	Use a non-glycerol-containing mountant
Nuclear counterstain is too heavy in the case of nuclear antigens	Counterstain for a shorter length of time and/or differentiate for longer in the case of regressive haematoxylin
Positive-control specimen demonstrates specific staining with minimal or very little background staining. Weak or absent immunochemical staining in test specimen, with varying degrees of background staining	
Antigen not present or present at too low a concentration in the test specimen	Use a more sensitive immunochemical detection technique (see p 63)
Test specimen is not adequately fixed	Fix for an extended length of time
	Fix smaller pieces of tissue to aid fixative penetration

TABLE 5.1 (*Continued*)

Test specimen fixed for excessive length of time in cross-linking fixative	Optimize and standardize fixation times
	Perform antigen retrieval
Test specimen exhibits necrotic/damaged areas or crush artefacts	Cut sections with a sharp microtome or cryostat blade
	Ignore such areas
General background staining demonstrated in both test and control specimens	
Excessive incubation times of reagents	Optimize and standardize incubation times especially primary antibody and chromogen
	Refer to product literature for suggested incubation times
Excessive reagent concentrations	Optimize working dilutions or concentrations for each reagent
	Observe manufacturer's recommended working dilutions or concentrations for commercial reagents
Inadequate rinsing of slides	Rinse slides adequately
Secondary antibody binding to endogenous immunoglobulins and/or antigens in specimen	Use a primary antibody raised in a species different from that of the test specimen
	Use a secondary antibody pre-adsorbed against normal serum or tissue extract from the species of the test specimen
Incorrect blocking serum used	Use serum from the species in which the secondary antibody was raised
Inadequate or absent blocking of endogenous peroxidases and/or phosphatases	Perform appropriate block(s) (see p 64)
Endogenous biotin when using an ABC immunochemical staining method (common in kidney and liver specimens)	Perform a biotin block using an excess of avidin followed by an excess of biotin. Use a proprietary blocking kit or see p 144
	Use a non-biotin immunochemical staining method
Drying of the specimen before fixation or during immunochemical staining	Immerse specimen into fixative immediately after being dissected
	Carry out all immunochemical staining incubations in a humidified environment
General background staining demonstrated in negative reagent control and test specimen. Positive- and negative-control specimens demonstrate expected staining with minimal or very little background staining	
Test specimen not adequately fixed	Fix for an extended length of time
	Fix smaller pieces of tissue to aid fixative penetration
Test specimen is fixed for excessive length of time in cross-linking fixative, increasing tissue hydrophobicity	Optimize and standardize fixation times
	Perform antigen retrieval
Test specimen exhibits necrotic or damaged areas or crush artefacts	Cut sections with a sharp microtome or cryostat blade
	Ignore such areas

TABLE 5.1 (*Continued*)

Excessive or patchy application of tissue adhesive to slides, therefore adhering immunochemical reagents	Observe manufacturer's recommended coating procedure when using commercial solutions
Tissue sections cut too thickly, 'trapping' immunochemical reagents	Aim for 4-μm thick sections and certainly no thicker than 10 μm
General background staining demonstrated in negative reagent control specimen. The positive-control, negative-control and test specimens demonstrate the expected staining with minimal or very little background staining	
Negative-control serum is too concentrated	Dilute the negative-control serum to the immunoglobulin concentration of the primary antibody (or whole protein concentration if the primary antibody is non-purified)
Antibodies present in the negative-control serum are cross-reacting with tissue components in the test specimen, or bacterial/fungal infection of the negative-control serum	Use a new batch of negative-control serum
Leucocyte staining is demonstrated in all or some specimens (usually frozen sections)	
Binding of the Fc region of the primary antibody to Fc receptors on leucocytes	Use Fab or F(ab)2 primary antibody fragments
	Add detergent to buffers to dissolve Fc receptors (see p 65)
Specific but unexpected staining is demonstrated in all or some specimens	
Contaminating antibodies are present in the polyclonal primary antibody solution that detect a different target to the desired one	Use an immunogen-affinity-purified version of the antibody (or at least a protein A or G (or protein Y for chicken antibodies)-affinity-purified version)
Positive staining on negative-control section	
Primary antibody has not been omitted from section used as the negative control	Ensure that the primary antibody has been omitted
Contamination of the secondary antibody during a previous procedure. Labelled antibody now binds to a tissue epitope	Ensure that a fresh pipette tip is used every time and never reuse reagents
Endogenous biotin. This can be a problem when using avidin–biotin detection systems on tissues such as kidney and breast	Perform a biotin block using an excess of avidin followed by an excess of biotin. Use a proprietary blocking kit or see p 144
Poor tissue morphology	
Loss of cell and tissue integrity, usually associated with poor fixation and paraffin processing	Reprocess the paraffin block or select another block on which to carry out the tests
Excessive proteolytic enzyme digestion or over-retrieval (HIER)	Check the concentration of enzyme and duration of digestion. Modify the HIER technique

In view of the often complex nature of immunochemical techniques and as a further validation of laboratory procedures, participation in an external quality assurance scheme is recommended. The UK National External Quality Assessment Scheme for Immunochemistry (UK NEQAS-ICC) assessments takes place at 3-month intervals throughout the fiscal year. Currently, laboratories are able to participate in up to six different modules depending on their service commitments and specialized areas of interest. These modules are as follows:

1. The general pathology module, catering for routine markers used by the majority of pathology departments offering a routine immunochemistry service.
2. The breast hormonal receptor module, catering for laboratories routinely demonstrating oestrogen and progesterone receptors on paraffin-embedded tissues.
3. The breast HER2 module, catering for laboratories routinely testing for HER2 on paraffin-processed tissues.
4. The lymphoma module, catering for laboratories with a special interest in lymphoma pathology.
5. The neuropathology module, catering for the markers common to most neuropathology laboratories.
6. The cytology module, catering for markers commonly requested on cytological preparations.
7. The alimentary tract pathology module, catering for laboratories specializing in this area of pathology.

More information can be obtained from http://www.ukneqas.org.uk/.

REFERENCES

1. Helander, K.G. (1994) Explains the methods of formalin fixation and the problems it causes with immunocytochemical investigations. *Biotechnic & Histochemistry*, **69**, 177–179.
2. Miller, R.T. (2001) Technical Immunohistochemistry: Achieving Reliability and Reproducibility of Immunostains. Society for Applied Immunohistochemistry Annual Meeting, 8 September, 2001, New York, USA. – Describes the effects of various commonly used decalcification agents on tissue antigens.
3. Athanasou, N.A., Quinn, J., Heryet, A. *et al.* (1987) Effect of decalcification agents on immunoreactivity of cellular antigens. *Journal of Clinical Pathology*, **40**, 874–878.
4. Matthews, J.B. and Mason, G.I. (1984) Influence of decalcifying agents on immunoreactivity of formalin-fixed, paraffin-embedded tissue. *Histochemical Journal*, **16**, 771–787.
5. Mukai, K., Yoshimura, S. and Anzai, M. (1986) Effects on decalcification on immunoperoxidase staining. *The American Journal of Surgical Pathology*, **10**, 413–419.
6. Huang, S.N., Minassian, H. and More, J.D. (1976) Describes the effect of proteolytic enzyme digestion techniques on paraffin embedded material. *Laboratory Investigation*, **35**, 383–390.
7. Brozman, M. (1980) Antigenicity restoration of formaldehyde-treated material with chymotrypsin. *Acta Histochemica*, **67**, 80–85.
8. Boenisch, T. (ed.) (2001) *Immunohistochemical Staining Methods Handbook*, 3rd edn, DakoCytomation, CA, USA.

9. Miller, R.T. and Kubier, P. (1997) Blocking of endogenous avidin-binding activity in immunohistochemistry: the use of egg white. *Applied Immunohistochemistry*, **5**, 63–66.
10. Deyl, Z., Macek, K., Adam, M. and Vancikova, O. (1980) Studies on the chemical nature of elastin fluorescence. *Biochimica et Biophysica Acta*, **625**, 248–254.
11. Cowen, T., Haven, A.T. and Burnstock, G. (1985) Pontamine sky blue: a counterstain for background autofluorescence and immunofluorescence histochemistry. *Histochemistry*, **82**, 205–208.
12. Schnell, S.A., Staines, W.A. and Wessendorf, M.W. (1999) Reduction of lipofuscin-like autofluorescence in fluorescently labeled tissue. *Journal of Histochemistry & Cytochemistry*, **47**, 719–730.
13. Yin, D. (1996) Biochemical basis of lipofuscin, ceroid, and age pigment-like fluorophores. *Free Radical Biology & Medicine*, **21**, 871–888.
14. Clancy, B. and Cauller, L.J. (1998) Reduction of background autofluorescence in brain sections following immersion in sodium borohydride. *Journal of Neuroscience Methods*, **83**, 97–102.
15. Bancroft, J. and Stevens, A. (eds) (1996) *Theory and Practice of Histological Techniques*, 4th edn, Churchill Livingstone, New York, USA.

CHAPTER SIX

6

Automated Immunochemistry

Emanuel Schenck[1] and Simon Renshaw[2]

[1]Medimmune LLC, Gaithersburg, MD, USA
[2]Abcam plc, Cambridge, UK

INTRODUCTION

Automation of immunochemical techniques in the laboratory has taken on considerable momentum since it first gained recognition in 1992 [1]. Continuous progress in this area is fuelled by several factors, including improved consistency and quality of automated results, a growing scarcity of highly trained technicians required for manual staining and the need for faster turnaround times in delivering results.

We have witnessed considerable technological progress in the last decade, and it is therefore appropriate to expect that automated technology will emerge as an enabling tool for various types of laboratories in the future. Modern automated immunochemical platforms are capable of taking over every task in the immunochemical laboratory, performing

---

Revised by Simon Renshaw.

Immunohistochemistry and Immunocytochemistry: Essential Methods, Second Edition. Edited by Simon Renshaw.

essential and tedious steps such as antigen retrieval or dewaxing of paraffin slides. Most scientists would agree that having this technology available in the immunochemical staining laboratory is a significant advantage.

Although the benefits of automation have been systematically evaluated and documented in a large number of scientific publications [2–4], the recent spread of automated systems into all areas of immunochemical practice indicates that manual methods have been displaced as the 'gold standard' for various immunochemical applications and assays. In the absence of detailed performance measures for the various systems, it is essential to evaluate the individual platforms in terms of overall quality of staining produced for various assays.

As there are quite a number of different automated stainers on the market, selection of the appropriate platform is not an easy task. Although similarity of automated protocols to manual methods may be important for some laboratories, it is probably wise not to regard this as the main criterion for selection of a platform. It can be expected that the technology will continue to provide solutions that go beyond the automated adaptation of manual methods. This chapter is intended to review the merits of the latest technological advances that enable faster throughput of immunochemical staining methods and to provide some guidance on the drawbacks associated with some of the current technologies and commercially available systems.

It must be noted that this chapter does not contain any protocols. This is due to the fact that automated immunochemical staining machines follow protocols that are identical to those employed in manual immunochemical staining methods, with the various steps being programmed individually or as a pre-set package by the operator before staining commences. Such protocols can be found in Chapter 3, 'Immunostaining Techniques'.

Defining the Needs

The task of choosing the optimal platform for a particular laboratory is simplified considerably if the needs are clearly defined in advance. Some of the relevant questions to review before selection are listed in the following:

- Is it essential for the operator to have a system that has the highest throughput (as measured in slides/hour) performing immunochemical staining in large batches using relatively few different protocols, or should the laboratory be equipped with a system that allows the highest degree of experimental flexibility to cope with numerous different protocols and modifications?
- Are additional system capabilities important, such as *in situ* hybridization, slide dewaxing and antigen retrieval?
- Is the aim to provide walkaway operation of immunochemical staining with the highest degree of automation and standardization in order to free up valuable time for the

technical staff to perform other tasks, or is the amount of 'hands-on' time (for protocol selection and modification, slide loading, etc.) required of the operator less important?
- Is the highest degree of standardization of the procedures desirable?
- Is the ability of the vendor to provide optimized reagents for all desired applications essential, or is the overall reagent cost a major factor for the day-to-day operation of the laboratory?

For many laboratories, it can be expected that any one platform will not fulfil all of the requirements. This often means that operators choose to purchase several platforms to cover the entire spectrum of tasks.

Automated systems are generally employed to increase throughput, decrease slide-to-slide variability and introduce a new level of reproducibility and standardization. Other desired endpoints include improved traceability and self-checking of critical steps.

Overview of Automated Platforms for Immunochemical Staining

Various types of automated slide-staining systems are currently on the market. Most systems that are currently installed in histology laboratories are used to perform only immunochemical staining. Other systems are used to perform both immunochemical staining and *in situ* hybridization.

The latter technique demands that the system provides both a means to heat slides to perform denaturation and hybridization steps and the ability to prevent sample degradation by desiccation. A number of automated immunochemical staining platforms can also be used to perform tinctorial stains. In the past, slide-to-slide variability has been an issue for tinctorial stains performed on automated immunochemical staining systems. Many vendors have now adapted their staining protocols to overcome these problems of consistency.

System Contrasts

There are considerable differences between the various automated systems with regard to the following:

- Overall configuration of the system.
- Reagent application method.
- The ability of the instrument to apply heat to individual slides within a narrow temperature range.

These and other parameters have to be evaluated before individual platforms are looked at in more detail.

METHODS AND APPROACHES

Overall Configuration of the System

Automated immunochemical stainers fall into two types of category with regard to their basic configuration: array stainers and rotary stainers. In array stainers, the slides are arranged in a matrix configuration of rows and columns. Figure 6.1 displays the Leica Biosystems BOND-III, an example of an array stainer.

FIGURE 6.1 Leica Biosystems BOND-III, An Example of An Array Stainer.
Source: Reproduced with permission of Leica Biosystems

One advantage of these stainers is that slides can be removed as staining procedures are completed. Some systems also offer continuous batch processing, allowing batches of slides to be stopped and started independently. Array stainers generally allow the operator to mimic manual procedure and offer the highest degree of experimental flexibility with regard to protocol design and the use of reagents from different vendors.

Rotary stainers, on the other hand, use circular arrangements of reagents and slides. A carousel is used to rotate reagents into position before they are applied to slides situated on a circular, stationary platform.

Although rotary stainers limit the operator's ability to modify staining protocols to some extent, they do allow simultaneous execution of very different protocols. In rotary stainers, access to slides during the staining procedure is generally not as easy as in array stainers. This could be perceived as a disadvantage in laboratories that require staggered workflows.

Staining platforms are designed to be either closed or open systems. Indeed, some staining platforms can be converted from closed to open systems at the request (and subsequent additional cost) of the end user. Closed systems restrict the operator's ability to use reagents from sources other than the platform's distributor. Depending on the overall use of the system within the laboratory, this can have important implications on overall running costs. The cost per slide can be several times that of a manual run. Open systems are more or less designed to allow the use of reagents obtained from sources other than the instrument's distributor. This will often result in greater experimental flexibility in designing the individual staining assays. On the other hand, selection of an open system may have implications for the overall hands-on time necessary to operate the platform, and standardization of the procedures may not always be optimal. Diagnostic laboratories that perform more routine immunochemical staining work prefer closed systems for their assays. The demand for open systems tends to come from research laboratories. It is advisable to evaluate individual platforms and to determine in detail whether they offer the desired experimental flexibility.

Reagent Application Method

Another aspect to consider with automated technologies is the means by which liquids are applied to and displaced from the slides. Different reagent delivery techniques can lead to different problems.

Some systems exploit capillarity forces to ensure the even spread of reagent solutions over the slide. In these systems, the slides are placed in a vertical position and a gap between the slide surface and a slide cover is used to draw reagents over the surface. Other systems employ flat immunochemical staining: slides are positioned horizontally in the incubator and reagents are delivered by syringe-like or probe-type delivery systems or through disposable pipette tips. Problems with capillarity include the improper filling and draining of the capillary gap and are often related to variations in gap sizes for different types of sections.

Difficulties such as these can be avoided if the sections are mounted close to the bottom of the slide.

Rotary stainers dispense reagents through syringe-like or printer cartridge-like dispensers. The reagent container is rotated into position above the slide, and the syringe-like system is used to dispense a predetermined volume of the reagent on to the slide.

Probe-type dispensers, which are similar to devices in chemical analysers, are also used in a number of different platforms. Reagent carryover can, however, pose a problem with these if the rinsing mechanisms are not adequate, or if the probe's shelf-life is exceeded and the 'non-stick' coating becomes compromised. Some probe-type systems allow precise definition of reagent volume and drop zone. The drop zone for these stainers can be the top, middle, or bottom of the slide or a combination of zones. The reagent volume can be adjusted precisely for each of the zones. Some platforms have unique features that provide a uniform distribution of reagents across the test specimen and help to prevent evaporation.

Yet another method of dispensing reagent on to a slide is to make use of disposable pipette tips, which have to be loaded and ejected by a robotic mechanism. Loss of tips during the staining procedures and other problems related to pipette loading and ejection can occur in these stainers.

Ability of the Instrument to Apply Heat to Individual Slides within a Narrow Temperature Range

Individual and other heating methods are features of some platforms, which enable heat-mediated or enzymatic antigen retrieval and dewaxing. As antigen retrieval is often the most critical step in the success of an immunochemical staining procedure, these systems provide a significant advantage over manual methods in terms of standardization. Slide heating is also a prerequisite for dual functionality of platforms as truly automated stainers for immunohistochemistry and *in situ* hybridization. A further advantage of this feature is that, by elevating the incubation temperature of primary antibodies, equilibrium can often be achieved without the need to use extended incubation at 4 °C.

Other Special Features

Other technologically significant aspects of automated platforms revolve around the software and peripherals supplied with the platforms. The software of modern immunochemical platforms is generally user friendly. However, configuration of the system, selection of the protocol, loading and so on in some of the less-advanced systems can cost considerable time. Many platforms can produce audit trails, which are increasingly important in the controlled environment of certain types of laboratories. The workflow in many histology laboratories tends to be more and more defined by automated labelling systems and many platforms follow this trend using bar-coded labels, which determine the staining procedures. The ability to run multiple staining units with only one computer may also be considered an advantage of some of the platforms.

System Running Costs

Another major consideration is the overall running cost (cost per slide) of the system. The difference in cost per slide between manual technique and automated systems varies significantly for the different platforms. For example, cost per slide is considerably higher for the so-called 'closed' systems requiring reagents provided by the distributor of the platform. Use of other so-called 'open' systems will often not result in a notable increase in cost per slide when compared with that of manual technique.

System Failure Safeguards

When you take the overall number of slides and cost of reagents for each individual run into consideration, it becomes important to eliminate any factors that might interfere with the ability of the system to complete the run. Many systems do not complete a run if the power is interrupted, making uninterrupted power supply (UPS) units essential to avoid this happening. Features designed to avoid human error can be considered a major advantage in some of the platforms. For example, special features of these platforms include safeguard mechanisms against reagent depletion by calculating the total amount of reagent required before a run is initiated. These systems will refuse to initiate a run if the reagent volume is not sufficient to complete all of the staining procedures, and it is therefore a common practice for technicians to overfill reagents slightly. Bar-coding plays a major role in this, as it also allows the system to flag missing reagents that are required to execute protocols defined for individual slides. However, it must be stated that the best safeguard of all is proper care and attention to detail by the technician when setting up an immunochemical staining run. When programming runs involve multiple protocols/reagents, technicians must pay particular attention to slide and reagent location, also ensuring that slides are loaded with the specimens facing the correct way up. When using open systems, such as when performing a manual immunochemical staining run, technicians must ensure that all reagents are made correctly in order to achieve optimal and accurate positive staining.

After completion of all staining procedures, desiccation of slides can become a problem with some stainers if the slides are left in the system for an extended period of time. This is avoided in platforms that use a liquid coverslip or that employ regular buffer washes until human intervention.

Table 6.1 lists the common automated immunochemical staining platforms available at the time of publication. Each individual laboratory must assess the features of each individual platform to assess which one best suits their requirements, as outlined in this chapter.

TABLE 6.1 Common Automated Immunochemical Staining Platforms Available at the Time of Publication

Vendor	Product Name
Amos Scientific	AIHS 660
Biocare Medical	Nemesis 7200
	Oncore
	IntelliPath
BioGenex Laboratories	i6000 series
	Xmatrix series
Dako	Omnis
	Autostainer Link 48
IHC World	MY2300
Leica Biosystems	Bond series
Thermo Scientific	Autostainer series
Ventana Medical Systems	Benchmark series

OTHER FORMS OF AUTOMATION

Automated Antigen Retrieval Systems

When automated immunochemical platforms do not have on-board antigen retrieval capabilities, the end user is left with two options: to either perform antigen retrieval by hand (see p 51) or to employ a stand-alone automated antigen retrieval system. The latter is by far the best option, at least as far as reproducibility of results and quality control is concerned, since being automated it removes a good proportion of variability that would otherwise be introduced by the operator.

Akin to automated immunochemical staining platforms, antigen retrieval systems come with a variety of options. Overall antigen retrieval times and slide capacity will vary from system to system. The systems themselves can be based on water bath, pressure cooker or microwave principles. All have at least one enclosed chamber for the retrieval to take place. Parameters such as temperature and retrieval duration are nearly always user definable. Some have the ability to hold the temperature of the retrieval solution to just under boiling point and to help prevent tissue section dissociation from the slide. All systems are designed to be used with buffers to facilitate antigen retrieval, and some have temperature ranges that allow enzyme solutions to be used. Operators often use their own antigen retrieval buffers or enzyme solutions but, as with automated immunochemical staining platforms, often manufacturers offer their own formulations, manufactured to strict quality control standards, further enhancing reproducibility of results. Dewaxing is

TABLE 6.2 Common Antigen Retrieval Systems Available at the Time of Publication

Vendor	Product Name
Biocare Medical	Decloaking chamber™ NxGEN
BioGenex	EZ-Retriever
DBiosys	Montage Opus
IHC World	IHC-Tek™ Epitope Retrieval Steamer Set
Lab Vision™	PT (Pre-Treatment) Module

another feature offered by some automated antigen retrieval systems, as is the ability to generate quality control reports.

Like with automated immunochemical staining platforms, each individual laboratory must assess their individual needs and purchase a suitable automated antigen retrieval system that meets their requirements.

For further information regarding antigen retrieval, please see p 51.

Table 6.2 lists the common antigen retrieval systems available at the time of publication. Each individual laboratory must assess the features of each in order to assess which one best suits their requirements, as outlined in this chapter.

Digital Pathology

At the time of publication of the second edition of this book, digital pathology is arguably the most exciting recent advance in the world of immunochemistry and histology. It is set to revolutionize the way in which data from tissue and cytological specimen slides are obtained, assessed, reported and archived.

Following are several separately definable elements that currently make up the 'Digital Pathology Environment' [5].

1. *Scan*

   Slides that have been either immunochemically or tinctorially stained are digitally captured using a high-resolution scanner. Scanner software has the ability to recognize areas of stained specimen in order to reduce scan time by not having to scan the whole area of the slide. A very quick initial scan of each slide using a low-resolution camera identifies the specimen location. Operator intervention is sometimes necessary at this stage, since a weakly stained specimen or additional artefact on the slide (air bubbles, dirt, etc.) can 'confuse' the scanner, leading to incomplete scanning. An added advantage of this initial tissue recognition phase is that it also helps to keep the file size of individual images to a minimum.

   Automated scanning at full resolution can then commence and obviously negate the need for someone to sit at a microscope and manually capture representative images,

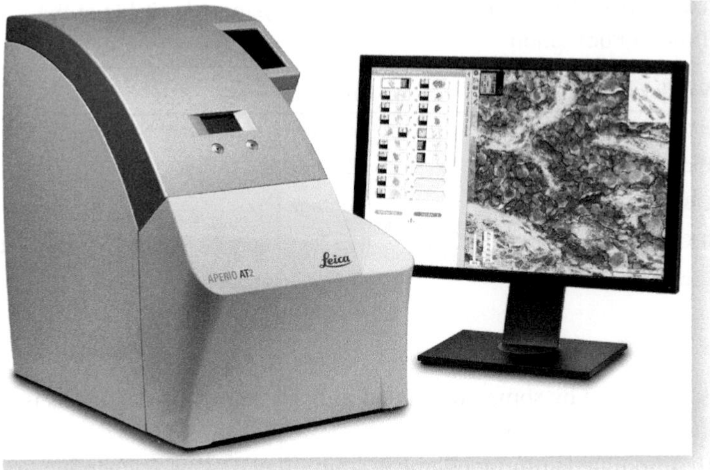

FIGURE 6.2 Leica Aperio AT2®, An Example of a High-Throughput Automated Slide Scanner.
Source: Reproduced with permission of Leica Biosystems

and since most scanners have high-throughput capability, an entire immunochemical run can be automatically scanned overnight, ready for analysis at the operator's convenience. Scanners also record and display the information on the slide label (usually at the initial tissue recognition phase) so that the subsequent results can be matched to the correct case.

Figure 6.2 displays the Leica Aperio AT2®, an example of a high-throughput automated slide scanner.

2. *Integrate and manage*
The entire digital pathology process can be tailored to fit seamlessly into a laboratory's existing workflow. The high-resolution images captured by the scanners are stored on a central database. Databases are security protected, backed up and archived. Depending on the amount of images generated, a substantial amount of storage space may be required, which is a major consideration when planning a digital pathology environment. However, the need for a potentially large amount of digital storage space is greatly offset by the fact that once the slides have been scanned, reported and backed up, they can be discarded, saving a considerable amount of physical storage space in the laboratory.
3. *View*
The images can be retrieved from the database by any PC workstation (or other compatible media device) with the necessary security permissions. This means that the operator can view the images from anywhere in the world. Images are typically associated with project or cases, and are grouped together as such, along with any other pertinent

information, such as patient information and history. Software packages to view images act like a virtual microscope: the operator can zoom in or out and scroll around to any area of the specimen at the click of a mouse button. Such systems are quicker and more pleasant to use than an actual microscope.

4. *Analyse*

   Analysis algorithms are frequently employed to enhance the amount of useful information that can be gleaned from a test specimen. They can range from simple cell counts to complex quantitative measurements, such as the percentage of positively stained cells in a particular area and breakdown of different staining intensities within those populations. Automated positivity scoring is made possible in this way. Spectral non-mixing (colour de-convolution) is a common analysis algorithm employed with multiple immunochemical staining, in order to effectively separate co-localized signals. Another very useful analysis algorithm is the de-arraying and subsequent analysis of individual tissue micro array (TMA) cores.

5. *Collaborate*

   As mentioned previously, the images can be retrieved from the database by any PC workstation (or other compatible media device) with the necessary security permissions. This means that anyone (not only the operator) can be granted permission to view the images from anywhere in the world, making it incredibly easy to share data when undertaking collaborative projects, or seek a second opinion on an outcome or diagnosis.

   Table 6.3 lists the higher profile digital pathology solutions available at the time of publication. Each individual laboratory must assess the features of each in order to assess which one best suits their requirements, as outlined in this chapter.

TABLE 6.3 Higher Profile Digital Pathology Solutions Available at the Time of Publication

Vendor	Product Name	Product Description
Leica Biosystems	Aperio® scanner series	Slide scanners
	eSlide Manager	Image management
	Aperio® ePathology	Image access and analysis software
	Aperio® PeerReviewer	Collaboration/image sharing software
	Digital SlideBox	Teaching software
PathXL	PathXL	Comprehensive image management and analysis software
Ventana	iScan Scanner series	Slide scanners
	Virtuoso	Image access software
	Companion Algorithm	Image analysis algorithms
	PathXchange	Online digital pathology community
	Ventana Vector	Interactive teaching and collaboration/image sharing software

REFERENCES

1. Grogan, T.M. (1992) *American Journal of Clinical Pathology*, **98** (Suppl. 1), S35–S38.
2. Moreau, A., Le Neel, T., Joubert, M. *et al.* (1998) *Clinica Chimica Acta*, **278**, 177–184.
3. Le Neel, T., Moreau, A., Laboisse, C. and Truchaud, A. (1998) *Clinica Chimica Acta*, **278**, 185–192.
4. Biesterfeld, S., Kraus, H.L., Reineke, T. *et al.* (2003) *Analytical and Quantitative Cytology and Histology*, **25**, 90–96.
5. Wikipedia (2015) *Digital Pathology*. [online], https://en.wikipedia.org/wiki/Digital_pathology (accessed 2 July 2015).

CHAPTER SEVEN

7

Confocal Microscopy

Ann Wheeler

Institute of Genetics and Molecular Medicine Advanced Imaging Resource,
University of Edinburgh, Edinburgh, UK

INTRODUCTION

Confocal microscopy is used to acquire a high-resolution image of a biological sample without any out-of-focus light. This is advantageous because it means that only the stained epitopes in a biological sample are seen without artefacts from background noise. This is particularly important for any biological specimen thicker than half a micron, as scattered and out-of-focus light pose a major challenge to collecting clear image data. In bioscience research, confocal microscopy allows in-depth study of tissues, cells and subcellular processes in both living and fixed tissues and can be used in a wide range of bioscience research areas. Confocal, as a tool, can be used for localization and co-localization of proteins, imaging multiple fluorescent stains, visualization of objects in 3D, semi-quantitative analysis of protein expression and live cell imaging.

In this chapter, we will discuss the following:

- When confocal microscopy should be used?
- How confocal microscopy works?

Immunohistochemistry and Immunocytochemistry: Essential Methods, Second Edition. Edited by Simon Renshaw.

- Best practice in setting up a confocal experiment
- How to set up a confocal microscope correctly for imaging multiple stains and co-localization experiments?

Later sections explain different applications of confocal microscopy, including tiling for histopathology, best practice in setting up a 3D confocal experiment, multi-photon confocal microscopy, live cell imaging (including emerging technologies) and quantitative image analysis.

History of Confocal

The first confocal microscope was developed in 1950 by Marvin Minsky. His design of using a pinhole to remove the out-of-focus light and retain the in-focus light is still used today in modern confocal microscopes. However, it took almost 40 years from the invention of the first laser scanning confocal before a commercially available confocal became available. In the past 20 years the technologies that support confocal microscopy have moved forwards in leaps and bounds. Major improvements in detector sensitivity, speed of data collection and computer technology mean that confocal microscopy can now be used for a wider range of scientific applications than earlier.

Widefield versus Confocal Microscopy

Widefield epifluorescent microscopy is the most straightforward of microscope set-ups. A sample is placed on a stage and the sample is illuminated from a broad-ranging, high-energy light source. Historically, this would be a mercury arc bulb, and several of today's systems still use this illumination source. Innovations in technology mean that the illumination sources for fluorescent microscopy are changing. Some systems are now illuminated by mercury halide or xenon bulbs. These light sources have the advantage that they last longer (around 2000 h) compared to mercury arc bulbs. The newest innovation in fluorescent illumination are LED light sources. These can last up to 20,000 h and are highly stable.

In fluorescent microscopy, light is split up by band-pass or long-pass excitation and emission filters. These may be in filter wheels or filter cubes. The filters select a specific range of wavelengths, which can be used to illuminate the sample. Fluorescent dyes will emit light at a longer wavelength than that of which they are excited by. The excitation and emission wavelengths are separated by dichroic mirrors.. Dichroic (or interference) filters selectively allow light of a certain range of wavelengths to pass and subsequently reflects light of other wavelengths.

In widefield epifluorescent microscopy, the whole of the field of view is illuminated. This means that the image will not only show the in-focus object but also light surrounding the object. This out-of-focus light can often mask the desired subject of study (see Fig. 7.1). For samples that are thin (around 500nm or less), very specifically labelled and not autofluorescent, this does not present a problem. Most biomedical research samples, however, are not like this. So to collect a high-quality image, it is necessary to remove the out-of-focus light,

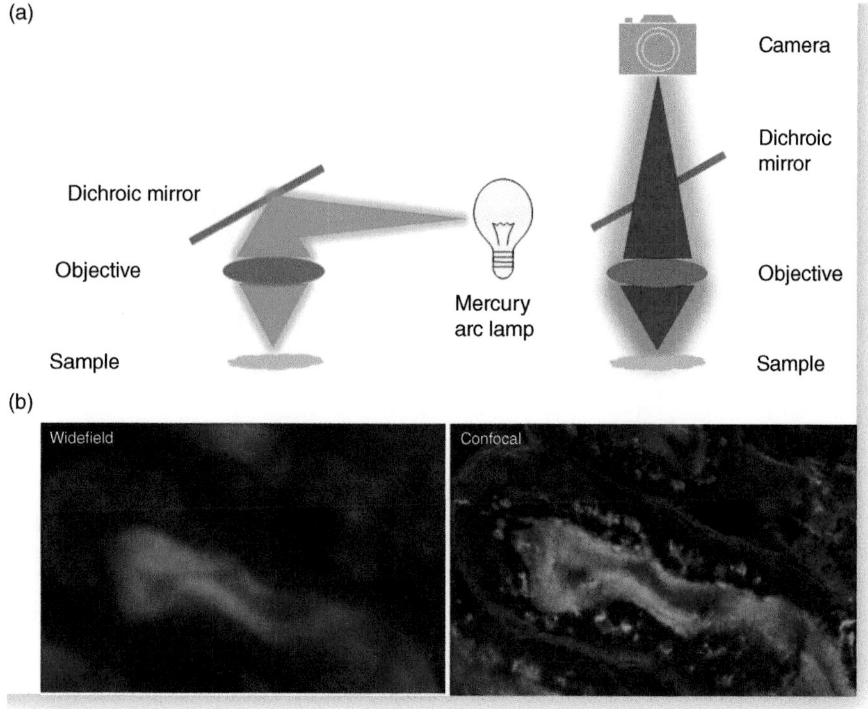

FIGURE 7.1 Widefield Epifluorescent Microscopy versus Confocal. (a) A Schematic Diagram of a Widefield Epifluorescent Microscope (Green for Excitation Wavelength of Light, Red for Emitted). (b) Shows the Comparison of a Widefield Epifluorescent with a Confocal Microscopy Image of Gut Epithelia Cells. Nuclei Are Shown in Blue (DAPI), Dividing Cells in Red (EdU) and Cytoskeleton in Green (Cytokeratin)

which means use of confocal microscopy. One word of warning here: generally confocal microscopes will come with an option to evaluate the sample by eye – down the eyepiece. The illumination to the eyepiece will be epifluorescent, as it would be very dangerous for the class 3B/4 lasers used for confocal microscopy to be looked at by eye. This can mean that the sample may look slightly different when confocal microscopy is used. Generally, the sample would be a bit dimmer because all the out-of-focus light that can be seen by the eye is excluded. The sample may also look slightly different because any out-of-focus light masking the epitope of interest will be removed.

How a Confocal Works?

The presence of out-of-focus light causes problems in imaging as it can mask the view required in the sample of interest. Marvin Minsky determined (theoretically) that the out-of-focus light could be removed by putting a small hole in the light path, which excluded all of the out-of-focus light. The most commonly used method for acquiring a

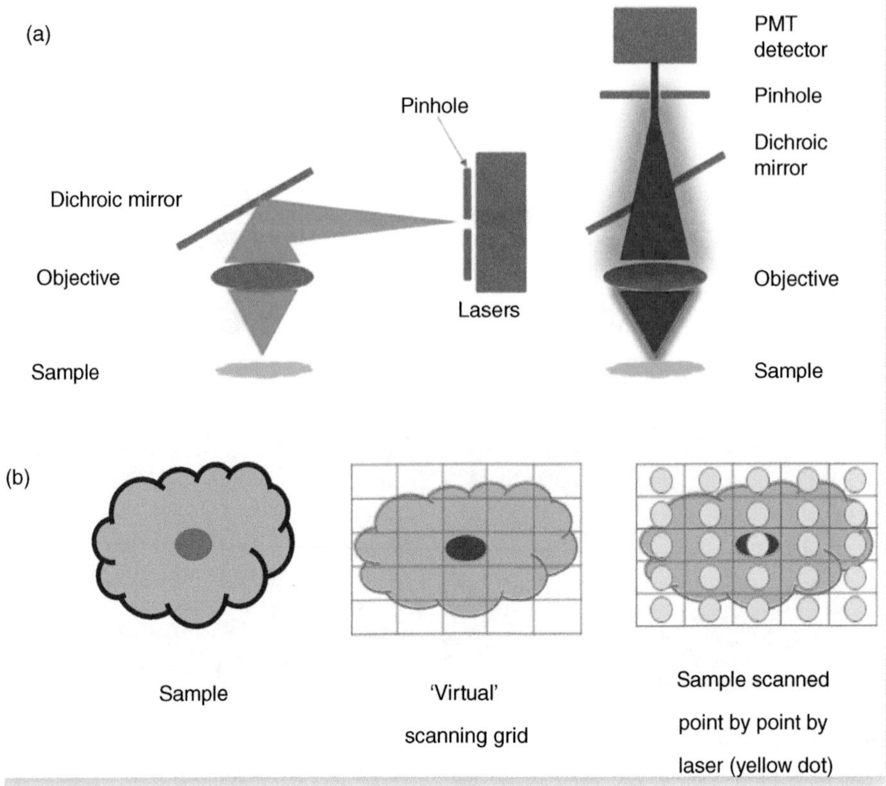

FIGURE 7.2 Schematic Diagram Detailing Confocal Set-Up: (a) A Simplified Diagram Showing Confocal Microscope Set-Up; (b) Schematic Diagram Showing Confocal Point-by-Point Scanning, Yellow Dots Indicate the Laser Line Scanning the Sample

point-scanned image is by putting a small hole (exactly the same size as the point spread function (PSF)) in the conjugate plane to the focal plane (i.e. the confocal plane, Fig. 7.2a). The small hole is called a pinhole. It is a necessity for pinholes to be adjustable in size because of the difference in the wavelength of light and the numeric apertures of the different microscope objectives used. In certain cases, microscopists may also choose to open up the pinhole slightly, allowing in a small amount of out-of-focus light to enhance the signal from the specimen.

In confocal microscopy, the pinhole is scanned across the specimen point by point, building up an image without any out-of-focus light present (Fig. 7.2b). This effectively means that the sample is divided up into a grid of small points that are visited in turn by the laser. The speed at which the laser visits each point, the dimensions of the grid and the frequency of times the same point in the grid is scanned can be varied. This will be discussed later. The point by point scanning of a sample means that a 3D focussed image of a specimen can be built up, without any out-of-focus light present. The confocal detectors are photo-multiplier tubes (PMTs). These are extremely sensitive. However, there is noise present in the detectors from heat, vibration and stray photons of light, which can make the output image look grainy. Therefore, to improve confocal imaging, it is suggested that a rough image is initially acquired to ensure the sample is in focus using a high scanning speed. Then the sample is scanned again, moving the pinhole across the specimen at a slower speed, to reduce the noise. The pinhole can also be moved to image an individual point a number of times, and by averaging the signal from that point an image that contains less noise can be obtained. This is called 'averaging'. To illustrate the benefit of using a slower scan speed and frame averaging, see Figure 7.7.

Most confocals are equipped with at least three PMTs. These are optimized for sensitivity in different parts of the spectrum. Some confocals are supplied with a spectral detector. This device can detect light from across the visible spectrum (400–750 nm). This spectral detector can greatly enhance the range of applications and the number of fluorophores that can be detected on a confocal. If imaging with more than three fluorophores is required, the imaging system may need to be set up to acquire channels in sequence that will take longer. Care must be taken here that there is no shift in the position of the fluorescent signal by 1 pixel when the configuration for the additional fluorophores is used.

WHEN SHOULD CONFOCAL BE USED?

Confocal microscopy can be used to visualize structures that are between 200 nm and 2 cm in size. Recent advances in stage automation have allowed for acquisition of larger images and this will be covered in the section about tiling large images for histopathology (see the following).

APPLICATIONS: FOR EXAMPLE CO-LOCALIZATION, QUANTIFICATION, 3D VISUALIZATION AND KINETICS

Samples such as cultured cells, tissue sections, sections of plant tissue and model organisms (e.g. fruit fly *Drosophila* and Zebrafish) are suitable for confocal microscopy across a very wide range of applications. The most commonly used application is the acquisition of a 2D image of a sample labelled with two or three fluorophores. These images can be used to study the distribution of a protein in a sample, localization and co-localization of two epitopes under different conditions. These images can be quantified to generate data about relative abundance of the fluorophores, spatially. Confocal microscopy can also be used to obtain data about multiple positions in a sample and live cell imaging. Since confocal microscopy excludes out-of-focus light from the image plane, it is possible to use confocal to build up a three-dimensional image of a sample, thus allowing a complete reconstruction of the spatial distribution of proteins in three dimensions.

Limitations of Confocal Microscopy: Depth Penetration and Resolution

The depth to which light can penetrate from a confocal is limited. Biological samples can absorb light, but they can also scatter it. The longer the wavelength of the light, the better absorbed it is by tissue samples. So dyes that absorb and emit light in the red or far red part of the spectrum are easier to image in tissue than dyes with shorter wavelengths.

As explained in the previous section, the confocal pinhole removes all of the out-of-focus light from an in-focus point on a sample. This means that it is possible to scan cells and tissue sections in three dimensions. However, the depth of penetration of light is limited by three factors:

1. The wavelength of the absorbed light (longer wavelength light penetrates further than short wavelength light, so red and far red dyes are better for imaging deep into a tissue).
2. Scattering of the excitation and emission light by the tissue or cells.
3. Photobleaching of the dye.

Taken together, these factors mean that for a standard biomedical research sample, it is only possible to image a few microns (6–10) into a sample. This means that thicker tissue samples will be sectioned, using a microtome before they can be imaged (see p 47).

HOW TO SET UP A CONFOCAL EXPERIMENT?

Visualization of the Sample on the Confocal

Good sample preparation is essential for confocal microscopy. Samples should be flat and carefully fluorescently stained. The fluorescent staining should be optimized to suit the detectors on the confocals. Older confocals (pre-2008) have less sensitive photo-multiplier tube detectors and so brighter fluorophores are preferred. However, the new generation of confocals use newer detector technology. These detectors are very sensitive, and so proteins that are less abundant (which would give a weaker fluorescent signal) can be detected. It is advisable to titrate both the primary and secondary antibodies prior to starting confocal imaging to ensure that the optimal signal is detected. For confocal microscopy, it is important to have the sample labelled with fluorescent dyes that have a high quantum yield. The quantum yield is a number that shows how many photons of light one will get out of a fluorescent dye with respect to the number of photons absorbed. The maximum this number can be is 1.0, as that would suggest for every individual photon of light which excites the dye one photon is emitted.

The cover glass on the top of the sample should be cleaned with water prior to imaging, as dried PBS can generate artefacts in the image and be a cause of auto-fluorescence.

Setting Up a Microscope for Confocal Microscopy

To understand how confocal microscopy works, it is necessary to understand a little of the physics of optics which underpin the basic design of a confocal microscope. If light transmitted from an object (such as a biological specimen) passes through a lens, it forms an image. The properties of that image, such as how well resolved (sharp) the image is, the colour preservation and how focused it is, greatly depends on the properties of the microscope objective lens. The microscope lens itself is labelled to give the end users a good idea of how it will perform in a given experiment (Figure 7.3).

When choosing the lens for your sample you will need appropriate magnification and good resolution. The detectors on any microscope will only take a square box in the centre of the field of view, so the area of the specimen which needs to be imaged should be placed in the centre of the field of view. Magnification and resolution are not the same thing. Higher magnification does not necessarily mean better resolution, as is shown in Figure 7.4.

(a)

(b)

Label	What it does
10 x, 20x, 40, 60x, 100x	Magnification of the lens
NA (numerical aperture) i.e. 0.3, 0.6 ,1.3	Numerical aperture, shows how much resolution you can expect
Acro, Fluar, Apo	Corrects chromatic aberration, makes sure all the wavelengths of light focus in the same place.
Plan	Ensures that the field of view is flat, that is in the same focus and lens curvature is accounted for
Ph DIC	Shows that the lens is also suitable for the specialist bright-field applications of phase contrast (Ph) and differential interference contrast (DIC).

FIGURE 7.3 (a) A Typical Objective Lens; (b) Table Explaining the Text Written on the Objective Lens

(a) (b)

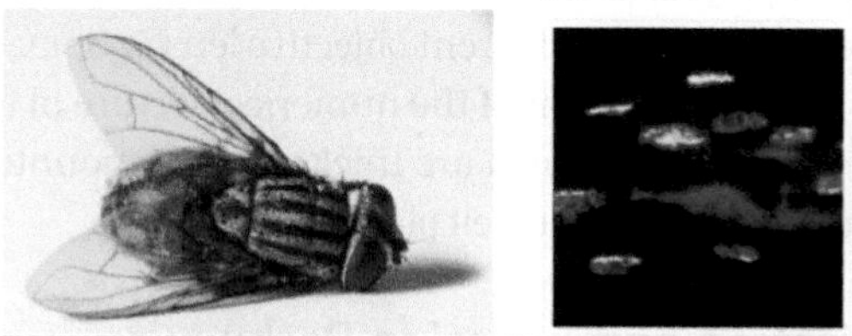

FIGURE 7.4 (a) A High-Resolution, Low-Magnification Image; (b) A High-Magnification, Low-Resolution Image

TABLE 7.1 Relationship between Maximum Resolution and Magnification, Numeric Aperture, Immersion Medium and Emission Wavelength

Medium	Magnification	Numeric Aperture (NA)	Thickness of Optical Slice (μm)		
			UV (DAPI)	Blue (GFP)	Green (Cy3)
Air	10×	0.3	9.6	11.5	14.0
Air	20×	0.5	3.4	4.1	5.0
Air	40×	0.7	1.5	1.8	2.1
Oil	40×	1.3	0.7	0.9	1.1
Oil	63×	1.4	0.6	0.7	0.9
Water	40×	0.8	1.7	2.1	2.6

The maximum resolution which an objective can give is defined by the wavelength of light and the size of the numerical aperture (NA). This was described in the 1800s by Ernst Abbe and is expressed by Abbe's law. The NA defines the range of angles of light which can pass through the objective and therefore determines how well resolved an image is (Table 7.1). Although the confocal will display an image of a section in two dimensions, the in-focus light collected by the objective lens will comprise of three dimensions. The thickness of the third dimension is described as the depth of field. For a 10× 0.3 NA objective, this is about 12 μm. However, for a 63× 1.4 NA oil immersion objective, the depth of field reduces to 0.3 μm (Table 7.1). At a certain point of magnification and resolution, Snell's law of refraction comes into play. This is because the sample to be imaged by confocal microscopy is generally placed under a thin glass coverslip. There is some refraction when light moves from the glass to the air. For low-power objectives with a low NA and a large depth of field, this is not noticeable. However, around 40× magnification, a considerable difference can be noticed (Table 7.1). Air objectives will often give fuzzy looking images because the depth of field is large and there is refraction of light in the sample. By selecting a lens with oil immersion, this can be overcome. The immersion oil supplied by manufacturers has the same refractive index as glass, which is why it is sometimes referred to as liquid cover glass. It is important to use immersion oil with the correct refractive index for the objective lens being employed, as using the wrong oil can generate refraction artefacts. Table 7.1 shows optical slice depth for different objective lenses. Some objective lenses have correction collars that allow adjustment of the numeric aperture of the objective that can be used to compensate for samples, which are thicker or are mounted on thick supports, for example tissue culture flasks or multiwell plates.

Selecting the Excitation and Emission Pathways

Confocal microscopes generally use lasers to excite fluorescent dyes and PMTs to collect the subsequent fluorescent emission. For some systems, there will be an array of lasers of

TABLE 7.2 Guide to Standard Laser and Emission Filter Configurations on Confocal Microscopes

Laser Line (Excitation; nm)	Emission Filter Settings before PMT	Dyes
405	420–450 nm band pass	DAPI, Hoechst, Alexa Fluor 405
488	505–530 nm band pass	FITC, GFP, Alexa Fluor 488, Cy2, DyLight 488, CellTracker and MitoTracker green
543	550–600 nm band pass	TRITC, Alexa Fluor 543, Alexa Fluor 555, Cy3, DyLight 550
561	570–620 nm band pass	Rhodamine, Texas Red, Alexa Fluor 568, Cy3.5, mOrange FP, CellTracker and MitoTracker orange
640	650 nm long-pass	Cy5, Alexa Fluor 647, DyLight 633, Draq5

differing excitation wavelengths. A standard set-up would be 405, 488, 561 and 640 nm. However, this can vary depending on the manufacturer. Some confocals are now equipped with white-light lasers, which means that it is possible to excite any fluorescent dye (Table 7.2). It is important to find out how the confocal will collect light emitted by the fluorescent dye. Before starting work, it is important to know how many detectors are available on the system and what the properties of the detectors are. Some confocals will come equipped with PMTs, which will be able to detect one dye at a time. Other machines may have spectral detectors. Spectral detectors are spectrophotometers and consist of an array of PMTs that can collect light across the visible spectrum (Fig. 7.5). It is possible to split up the detection on the spectral detector, so as many as 34 different signals can be detected at once. It can be useful to understand a little about the path that the light will take through the confocal. In older machines (pre-2005), this was mainly controlled by fluorescent filter cubes. More modern technology makes use of Acousto-Optical Beam Splitters, diffraction gratings and prisms to separate light more efficiently. However, there will be limitations on the pathways that can be used to direct light from the laser to the detector in any confocal, which will relate to the beamsplitters, filters used and the performance of the detectors. These will limit the choice of fluorescent dyes that can be used and also the combinations of dyes that can be used together. The manufacturer of the confocal will be well placed to advise about which dyes are best for the machine. However, a general guide is shown in Table 7.2.

Multicolour Confocal Imaging for Co-localization

Most confocal microscope images comprise more than one fluorescent label. Ideally, the different fluorescent dyes will be excited by different lasers on the confocal. A typical dye set-up may be DAPI, a protein labelled with an antibody conjugated to Alexa Fluor 488 and a protein labelled with an antibody conjugated to Alexa Fluor 568. It is very important to ensure that the fluorescent dyes are chosen, which will work optimally with the confocal

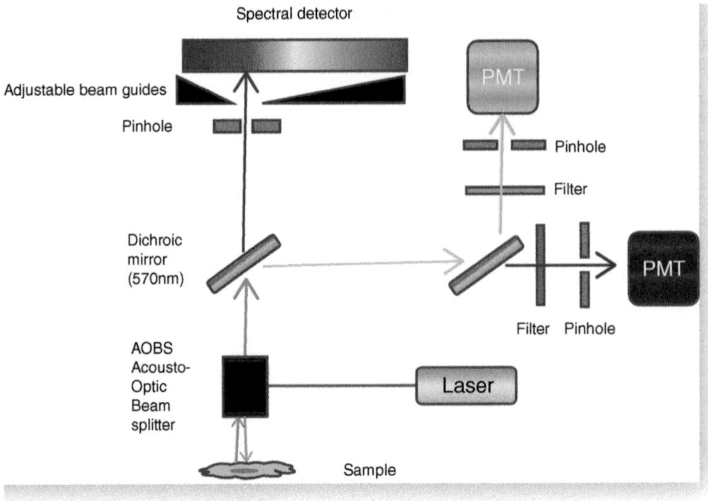

FIGURE 7.5 Schematic Diagram Showing Confocal Configuration of a Confocal with a Spectral Detector Which Can Detect Emission across the Visible Spectrum

instrument used. It is also important to understand the quantum yield of the dye that is to be used, in order to give an indication of how bright the dye is likely to be. Quantum yield is defined as the number of photons emitted for a given dye molecule per excitation photon, the brighter the dye is the more photons will be released per photon absorbed. Since we know that confocal microscopes will exclude a lot of light from the sample, it is best practice to choose dyes for confocal microscopy that have high quantum yields, reserving the dye with the highest quantum yield for the least abundant antigen. Novel Alexa Fluor, DyLight and Cy dyes all have high quantum yields and are excellent choices for confocal microscopy.

This does mean that some dyes that are used in FACS experiments may not be suitable, particularly if their excitation and emission spectra are not compatible with the filter-wheel settings of confocal microscopes. The dye spectra can be checked using online tools such as SpectraViewer: http://www.lifetechnologies.com/it/en/home/life-science/cell-analysis/labeling-chemistry/fluorescence-spectraviewer.html.

Spectral Bleedthrough (Cross-Talk)

The excitation and emission spectra of some fluorescent dyes will overlap. It is important, particularly for co-localization experiments, to try and select pairs of dyes where there is little spectral overlap. This is because if the emission spectra of one dye bleeds into the other channel, it can lead to false-positive detection of signal and misinterpretation of data. In Figure 7.6, it can be seen that the red channel is bleeding into the green channel very strongly. This gives the impression that the red staining is found throughout the cells that are labelled. However, careful separation of the emission spectra using band-pass filters shows that this is clearly not the case. The red dye is only localized in the cytoplasm of the cells and not in the plasma membrane.

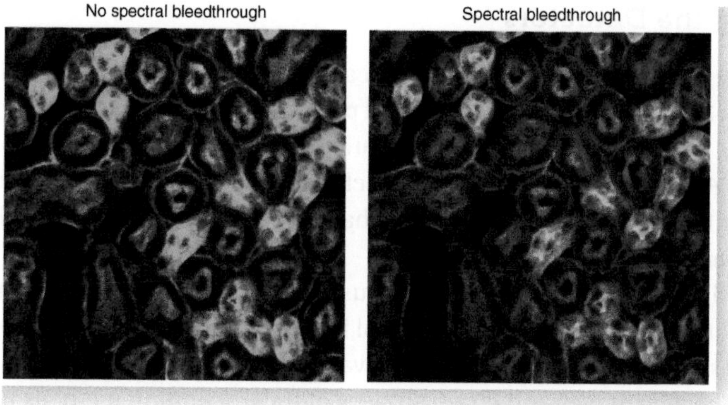

FIGURE 7.6 Spectral Bleedthrough Can Lead to Misinterpretation of Scientific Data. Confocal Images of Kidney Epithelia Labelled with *N*-Cadherin and CellTracker Orange

Best Practice in Setting Up Multicolour Experiments

When setting up a confocal experiment, it is not advisable to use the combination of Alexa Fluor 488 and GFP because they have very similar excitation and emission spectra and it is almost impossible to spectrally separate them out. A combination of Alexa Fluor 488 and Alexa Fluor 647, where there is very little spectral overlap, is strongly preferable. Selecting dyes with different excitation maxima is the best way to carry out a confocal experiment with multiple labels, for instance, a dye that is maximally excited around 490 nm and a second dye that is maximally excited around 570 nm. The fluorescent light emitted from these dyes can be separated using the band-pass filters and adjustable beam guides. Using filters, sequential scanning and careful selection of the sensitivity of the detectors, it is possible to minimize spectral bleedthrough. This is done by ensuring that the dichroic mirrors in the beam path are set up to separate out the light from each fluorophore.

To optimize the detectors for muticolour imaging, the best practice is to generate controls for fluorescent staining. Recommended controls are as follows:

1. Unlabelled sample (to obtain an image of background or auto-fluorescence in the sample: flavins such as FAD and NADH are highly fluorescent and can cause unwanted background).
2. A sample labelled with each of the fluorophores separately.

The confocal is then configured so that each channel detecting an individual fluorophore is only collecting that one specific signal, and there is none of that signal detected in any of the other channels.

For further information on spectral bleedthrough, please see *Basic Methods in Microscopy – Spector and Goldman* [1] or the molecular expressions website http://micro.magnet.fsu.edu/ (see [2]).

Optimizing the Detectors

Once the sample has been put onto the confocal and focused by eye, the light is sent to the confocal detectors. First of all, the correct light path has to be selected on the confocal. Some machines will come with premade configurations for frequently used dye combinations, for example DAPI, FITC and TRITC. Other machines will need to be configured by the user. Generally, confocal microscope software will have a way of saving a given configuration of lasers, filters and detectors for later use.

Initially, the light path to each detector must be set up according to the manufacturer's instructions. It is possible to scan the confocal image with all of the lasers simultaneously, or each laser sequentially, with the laser-activated changing after either a line or a frame is scanned. The advantages and disadvantages of these scanning methods are shown in Figure 7.7. As can be seen, simultaneous scanning is quicker but there are issues with spectral bleedthrough in the sample. Sequential scanning is much slower, however, because each laser is activated separately, there is a considerably reduced likelihood of bleedthrough occurring. Therefore, for imaging of fixed tissue, acquiring confocal images by sequential scanning is best practice. It is possible to set up an imaging experiment, where two channels which are spectrally very far apart, for example DAPI and Alexa Fluor 647, are scanned simultaneously to save time. Care needs to be taken setting up these experiments so that there is no cross-talk. On most confocals, it is possible to sequentially scan line by line or frame by frame. Some users will find scanning sequentially line by line most helpful as it assists with focusing the sample for all of the dyes used. These scanning methods can be set up in the confocal software and saved for later.

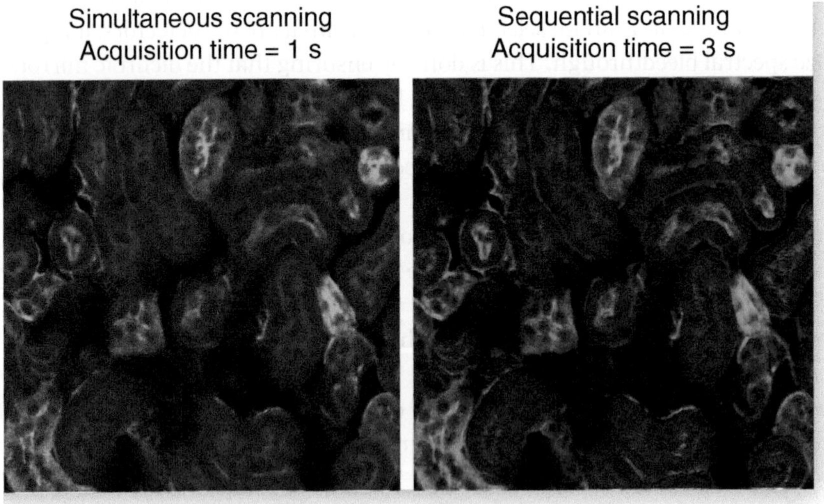

FIGURE 7.7 Comparison of Simultaneous and Sequential Imaging for Confocal Data Acquisition

Optimizing the Confocal Image under Exposure and Saturation of Sample

To obtain the optimal image on the confocal, the first step is to focus the image on the detectors. To do this, it is preferable to set the confocal to a fast scan speed to ensure that the region of interest is not bleached. The first time the confocal is set up, it is advisable to generate the initial settings in part of the sample, which is not the region of interest. Optimizing confocal acquisition settings can take some time, and exposing the sample to intense laser light from the confocal may cause photodamage or bleaching of the sample.

Confocal PMTs, such as any detector, have a range within which they will work. It is important to include all of the information from the sample in the captured image. This includes the background. It is a good scientific practice to have the background of the sample set to the minimum that the PMT can detect and the brightest part of the staining set to the maximum intensity the PMT can detect. Most confocals will have a mode where the detector can be set to show saturated pixels and under-exposed pixels. The gain and the offset settings of the confocal can then be adjusted to ensure that there are no saturated or under-exposed pixels in the image (see Fig. 7.8).

It is possible to alter a function called bit depth, which controls the range of bins the intensity information is divided into, ranging from 0 which is the bottom of the range (or black) and white. For an 8-bit image, this is 255 – so there are 256 intensity bins. For a 12-bit image, there 4096 intensity gradations, and for a 16-bit image, there are 65,536 gradations. The more gradations in an image, the more quantitative information will be in the image. If the confocal image is being captured for the purpose of extracting quantitative data (e.g. measuring the amount of protein in a given cellular location) the higher the bit

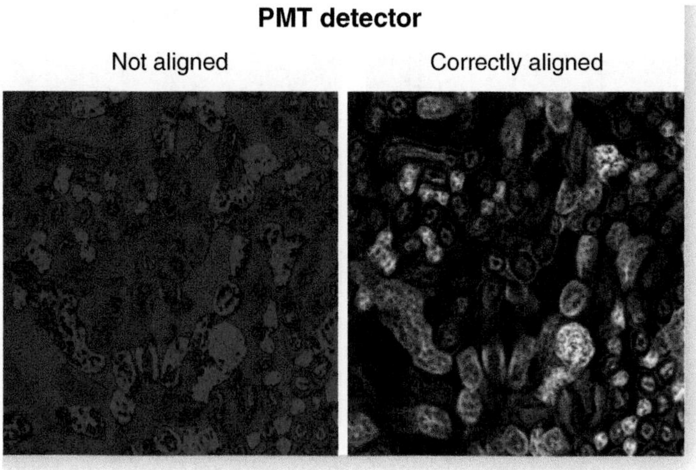

FIGURE 7.8 Best Practice in Alignment of PMTs. Confocal Image of N-Cadherin in Kidney Cells. Red, Saturated Pixels; Blue, Under-Exposed Pixels; and Grey, Pixels in Linear Detector Range

depth the image is acquired to the better. However, if an image is being captured for documentation or display purposes, an 8-bit image is sufficient. This is because the human eye can detect fewer than 250 tones of an individual colour so that an 8-bit image has plenty of detail in it for visualizing by eye.

Averaging of Scanning to Reduce Noise

When acquiring confocal images, the speed of the laser scanning can be varied. The slower it is, the less noise is present. There is a trade-off because the longer the laser dwells on an individual pixel, the more likely the specimen is to become photobleached. Apart from slowing the laser speed down, it is also possible to make the laser scan each pixel of the image more than once and produce the average. This averaging reduces noise and increases signal. In general, a pixel dwell time of 2–3 s and frame averaging of 4 will produce a high-quality confocal image. By slowing down the laser scan speed and increasing the frame averaging, the noise that is present in a confocal image can be reduced considerably (Fig. 7.9).

Changing the Size, Shape and Zoom of Sample Detected

Many confocals will have the ability to digitally zoom into a region of interest and to select the region of interest to be scanned. This added digital zoom can be advantageous, as it allows a more detailed area of a specific region to be recorded. This can be particularly helpful if an inset of the image is needed for publication or to illustrate a particular scientific

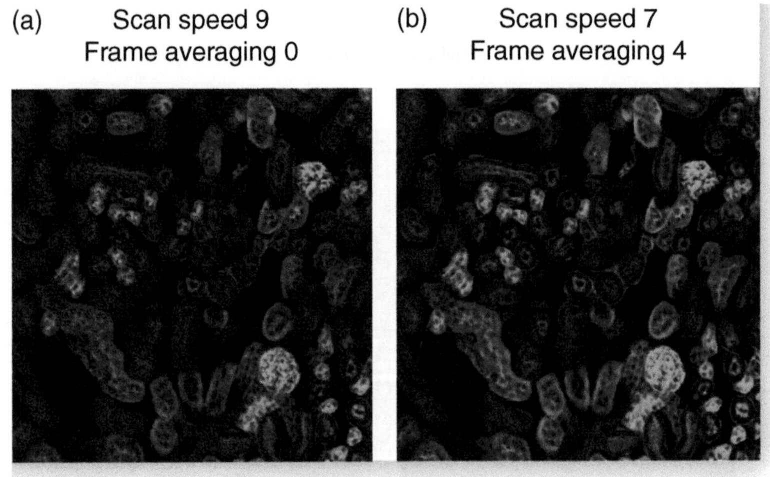

FIGURE 7.9 Scan Averaging and Altering Scan Speed/Pixel Dwell Time Improve Confocal Image Quality: (a) Image of Kidney Epithelia Was Scanned at High-Speed/Short-Pixel Dwell Time; (b) the Image Was Acquired with Longer Pixel Dwell Time/Slow Scan Speed and Displays the Average of Four mages

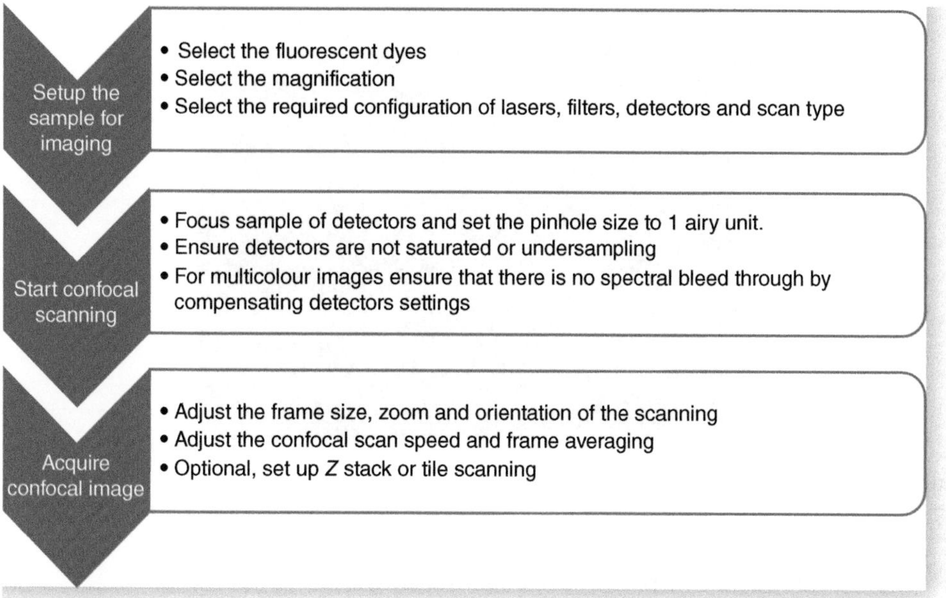

FIGURE 7.10 Confocal Microscopy Workflow Summary Diagram

point. It is also useful as specific areas can be scanned at slightly higher magnification than could be achieved by the objective lens alone. Increasing the digital zoom on the confocal image will not increase the overall resolution, instead it alters the size of the pixels on the instrument, that is would make the laser scan points that are closer together. This has the effect of enlarging the image, but can lead to the image being oversampled and yield pixelated data. As a rule of thumb, the effective pixel size in the scanned confocal image should be around twice the resolution limit. Figure 7.10 summarizes the discussion in the previous sections.

Three-Dimensional (3D) Confocal Imaging of a Volume

One of the major advantages of confocal microscopy is that the exclusion of out-of-focus light from the image allows the rendering of highly detailed 3D images, which have almost no out-of-focus light present. This allows visualization of a specimen in unprecedented detail and allows precise spatial mapping of multiple epitopes in cells, tissues and organs (Fig. 7.11). Not only are these images very beautiful to look at but they contain a considerable amount of useful biological information. Studies ranging from the size of telomeres in the nucleus to development of frog and fly embryos have been conducted using 3D confocal imaging. Indeed, with advances in detector sensitivity (hybrid detectors compared to traditional PMTs) and scanning speed of confocal microscopes, the fastest, most sensitive and most efficient way of acquiring data about biological volumes between 2 cm and 200 nm is by using confocal microscopy.

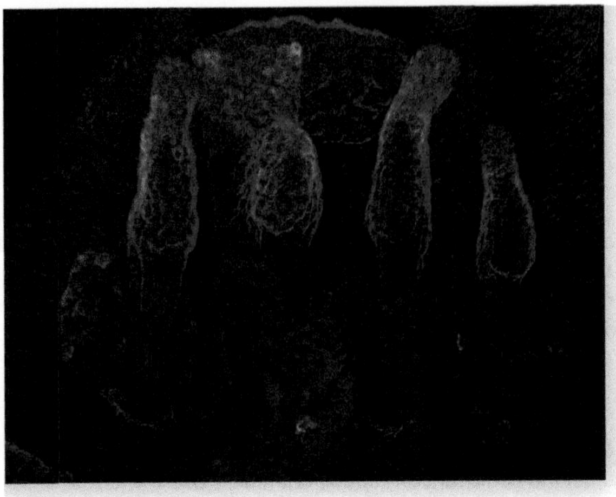

FIGURE 7.11 Three-Dimensional Confocal Image of Mouse Hair Follicles

How to Correctly Set Up 3D Confocal Imaging?

Confocal microscopy, like all forms of light microscopy, is limited by the diffraction capabilities of the optical parts of the microscope system. To understand how to set up 3D confocal microscopy, it is essential to understand a little of the physics which underpin microscopy.

The limit of resolution describes the distance between two points, which can be separately distinguished by a microscope in 3D space. In order to set up 3D confocal images, it is important to understand the limit of resolution of the microscope to ensure that specimens are sampled in such a way as to collect all of the 3D information present. Generally, for collecting confocal images, this means that they should be sampled according to the Nyquist theorem. Nysquist states that a sample should be oversampled by a factor of 2 to ensure all of the information is collected. So, to collect all of the information in a 3D confocal image, a biological sample should be collected at twice the spatial resolution that the microscope allows in all three dimensions.

To do this, it is essential to establish what the resolution of the microscope is. Each of the lenses, mirrors and filters inside the confocal microscope has limited efficiency in transmitting the rays of light. In some cases, they will also distort the light beam slightly (Fig. 7.12a). This can be measured by using a sample that is a small spherical point and assessing the properties of image of this small point generated by the microscope. The output image of the small point can be represented by a mathematical function, which is termed the point spread function (Fig. 7.12a). There is always more distortion axially (z-axis) of a point than laterally (x- and y-axis) (Fig. 7.12b). So a point of light will be converted into an ovoid shape by all microscopes (Fig. 7.12a), and confocal microscopes resolve less well axially than laterally. The resolution limit is often considered the full width at half maximal intensity of the image of the point source of light in the x-, y- and z-axis (Fig. 7.12b). It is possible to measure this for each lens and laser used on a confocal microscope. Then, once this is known,

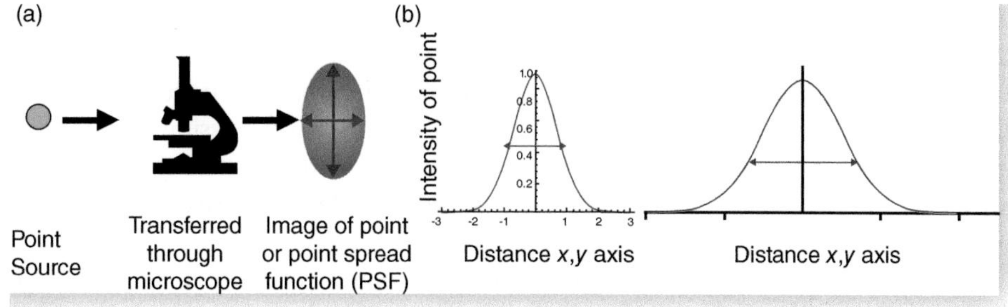

FIGURE 7.12 Resolution Limit in 3D of Microscope: (a) Schematic Diagram Showing How a Point Source of Light Is Distorted by the Optical Elements in a Microscope to Yield a Measureable Point Spread Function. (b) Measuring the Resolution by Looking at the Full Width at Half Maximal Intensity of Image of the Point Source Shows That the Resolution in *x,y* Is Half the Size in *z*

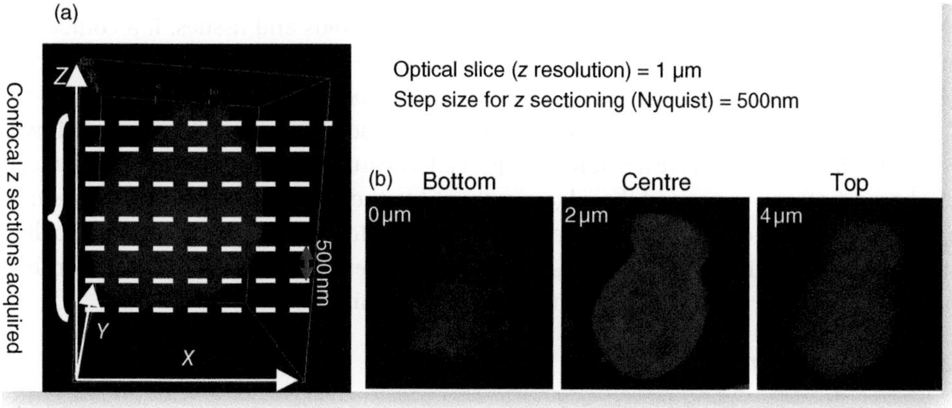

FIGURE 7.13 Acquiring *z* Stacks: (a) Diagram of Nyquist Imaging of a Nucleus in 3D with 1-um Optical Sections Using 500 nm Steps (b) Indicated How to Set Up a *z* Stack Where the Bottom and Top of the Object Are Defined and Then Imaged

the confocal microscope can be programmed to scan the image in *z* by moving the stage up in increments, which are half the size of the resolution limit: for 20× 0.5 NA lenses, the resolution in *z* would be 1 µm, so for Nyquist sampling 500-nm step sizes would be needed (Fig. 7.13a). Since confocal microscopes have several different lenses and at least three laser lines, most manufacturers have already done measured resolution for each lens and laser line and programmed it into the imaging software.

When setting up the confocal for collecting *z* stacks, there are several options available. For fixed imaging, it is preferable to define the top and the bottom of the sample. For a perfect image, Nyquist sampling in the *z*-axis should be chosen. Figure 7.13 shows an example of collecting the 3D volume of a nucleus using Nyquist sampling. Confocal microscopes are generally equipped with an automatic *z*-drive, and this can usually be programmed to step at user-defined intervals. It is possible to undersample in the *z* to speed data collection (as point scanning of volumes can be slow), but the resulting 3D reconstruction of the image will not be smooth and high quality (Figs 7.11 and 7.13) due to the missing *z* information. To reconstruct images in 3D, it is recommended to capture all of the structure/sample and have a few planes above or below the sample that are out of focus. The extra image planes can be useful if post-capture processing is required (Fig. 7.13b).

For live imaging of samples (which are expected to grow or move during the experiment), a different strategy may be needed, where either the bottom or the centre of the sample is defined, but it is not possible to define the top of the sample as it has not grown or moved there, yet. In such cases, the instrument can be asked to sample a user-defined amount of space above the bottom or ranging around the centre.

Collecting Larger Samples Using Confocal Tiling

The confocal volume that can be sampled is limited by the *x*- and *y*-dimensions of the detector. For larger samples, such as model organisms, organoids and tissues, if a confocal is equipped with a scanning *x*, *y*, *z* stages, the stage can be moved to acquire adjacent fields of view creating a tiled image (Fig. 7.14). It is very important that the sample is flat for this, and the optics of the confocal are well aligned. If there is uneven illumination, it will cause the final image to have a patchwork quilt effect caused by vignetting (reduced image brightness/saturation around the edges compared to the middle). For good tiling, a small amount of the edge of each tile will need to be blended into its neighbour in a process called stitching and blending. Several microscope manufacturers make good algorithms for this; however, open-source options in Fiji/ImageJ are also available: http://fiji.sc/Fiji.

Live Cell Imaging

Imaging live samples by confocal can generate extremely high-quality images, particularly in 3D. Three-dimensional confocal imaging has been used extensively in developmental

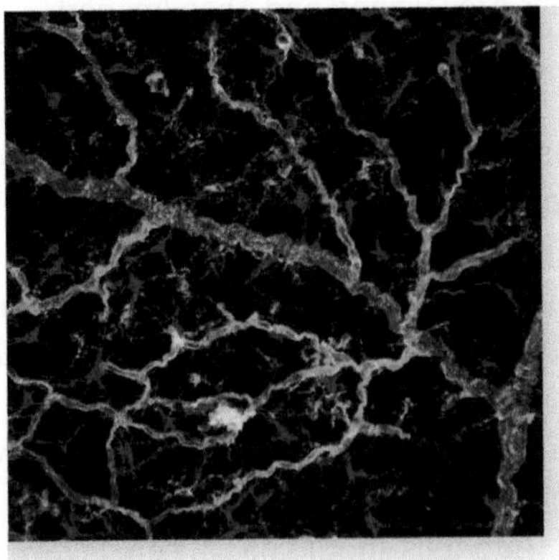

FIGURE 7.14 Tiled Image of Mouse Retina Olympus BioScapes 2013

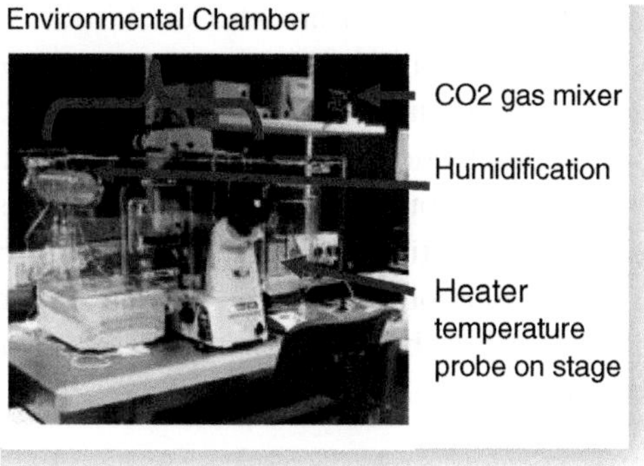

FIGURE 7.15 Confocal Microscope with Environmental Incubation for Live Imaging

biology for detailed studies of organ development. Model organisms such as *Xenopus Laevis*, *Drosophila* and Zebrafish are particularly useful models since they can fit into a field of view and are small/transparent enough to be visualized using confocal microscopy and survive. Live confocal imaging can also been used in mammalian models, but this requires some kind of environmental control to be added to the instrument. For mammalian samples, an environment similar to that of a tissue culture incubator must be generated (Fig. 7.15). This is often in the form of a large Perspex chamber, which is mounted onto the microscope that contains a heater and a thermostat to maintain the system at 37 °C. The thermostat should be placed as close to the sample as is possible, usually on the stage. Often, it is either mounted on or next to the stage. The sample will need to be humidified, so it does not dry out during imaging. Double-distilled water is used for this and is placed either in an adjacent well to the sample (if one is available) or in small containers next to the sample (Fig. 7.15). The plastic boxes that glass coverslips are supplied in work well for this. The pH of the medium that the sample is growing in should be maintained, and there are several schools of thought about this. The interested reader is encouraged to use the literature relevant to their field for expected standards. Either buffered medium such as Optimem (Life technologies) or Hepes, or direct supply of heated and humidified CO_2 can be used. Additionally, for live cell imaging of fluorescent dyes or proteins, it is necessary to omit Phenol Red from the medium as can auto-fluoresce and mask the desired signal. DMEM and F12 medium, which are phenol red free, are available. Evrogen supplys medium that is specifically generated for imaging with fluorescent proteins (DMEM-GFP). Some groups add Oxyrase to medium and others generate in house medium for live imaging dependent on the application.

In general, for live cell imaging, an inverted microscope stand is needed as samples are prepared in either glass-bottomed dishes or multiwell chambers. For confocal imaging, it is essential that aberrations due to refraction are minimized, so using a system where the refraction is matched through the objective lens, immersion oil and cover glass is ideal. For instance, high-resolution studies often use glass coverslips that are numbered as 1.5 with high-quality, fluorescence-free microscope immersion oil. For thicker samples, where light needs to penetrate deeper into the sample, specialist water immersion objective lenses with thinner coverslips (number 1) are also used. Intra-vital microscopy, using live model organisms such as mice, utilizes upright microscopes with very high-quality, water-dipping objectives, but these can only be used when they are fully immersed in aqueous medium and must be cleaned carefully as the salts in culture medium can corrode objective lenses. For more complex samples or sensitive samples that can only grow on tissue culture plates, objective lenses with adjustable numeric apertures are available. These are termed correction collars and can be set for a variety of different immersion media or samples. All of the large microscope manufacturers such as Leica, Nikon, Olympus, Zeiss and so on manufacture these.

Pros and Cons of Point Scanning for Live Imaging

Since fixed samples are, by their very nature, dead, it matters slightly less if high laser power or multiple scans (which are time consuming) are used. Live imaging of a sample using confocal microscopy presents a new set of challenges:

- Laser light is intense and can kill or compromise samples
- Scanning time needs to be traded off with required speed of acquisition
- 3D imaging can take a long time with point scanning
- Use of fluorescent probes is required and these can be bulky, highly charged and generate oxygen radicals that can perturb the biology. It can also be challenging to fluorescently label samples.

All of these need to be carefully considered when setting up a live confocal experiment. The consideration is that the sample has to be healthy while it is being imaged for any correct scientific interpretations to be made. The literature is littered with examples of poor confocal practice where high laser power or slow acquisition time has caused the samples to become unhealthy. This makes live imaging a complex art as the manual dexterity and judgement of the system operator can be key to obtaining good results. There are several excellent live imaging courses that are recommended to learn the pre-requisite skills. Generally, a trade-off needs to be made. By slightly compromising on image quality, number of sample points collected or *z*-resolution, it can be possible to collect live images which contain data that supports fixed images.

For visualization of samples for live confocal microscopy, either fluorescent dyes which can be incorporated into the sample or fluorescent proteins can be used.

Dyes: Most fluorescent dyes can be either transfected or taken up by the cells using endocytosis or pinocytosis. Many dyes contain an AM-ester moiety, for example the calcium indicators Fluo4 or Rhod3. This small chemical group allows a dye to permeate through the cell's plasma membrane and then be cleaved off once inside the cell. Several small molecules, such as Hoechst, Draq5, CFSE and others are also cell permeant. Loading of these dyes needs to be optimized for the individual microscope and biological system. Too much can compromise the biological system or oversaturate the detector, whereas not enough will either be undetectable or require such a high laser power that the sample is compromised by the imaging parameters required. It is often necessary to wash off the loading dye as the background can mask the sample. It is essential to check that the excitation and emission spectrum and emission spectrum of the dye is compatible with the confocal microscope system before work is begun. Some chemical probes such as Fura2 require long-wave UV light (350 nm) for excitation which may not be available on the microscope system. Other probes such as Lucifer Yellow or BCECF have excitation and emission spectra which are different from standard fluorescent dyes such as Alexa Fluor488/GFP. To image, these may require specific configuration of the confocal system. For systems equipped with a tunable emission filter/spectral detector (e.g. Zeiss LSM 710, 780 and 880; Leica SP5, SP8), this does not present a problem as the emission can be custom-tuned. However, set-ups which are filter-cube based (Nikon A1, Zeiss LSM 510, Leica SP2) may not be able to image these dyes unless the correct emission filter-cube is present.

Fluorescent proteins: There are an ever-increasing array of fluorescent proteins available, meaning that cloning and transfection are the only barriers to live imaging of many proteins. Brightness, as described by quantum yield, is important. Older red variant fluorescent proteins such as RFP and mCherry have low quantum yields and are likely to photobleach unless sensitive set-up and care are used. More recent red fluorescent protein variants such as mOrange and tagRFP suffer less from this problem. A comprehensive list of fluorescent proteins currently available is available here: http://nic.ucsf.edu/dokuwiki/doku.php?id=fluorescent_proteins.

It is worth mentioning that on systems equipped with argon lasers which have four lines, the 488-nm line is often considerably stronger than the 458 line, which is used to excite CFP. A laser power metre can be used to measure the power of the sample to check this. Care should be taken when generating or using fluorescent-tagged proteins. If possible, monomeric fluorescent proteins are preferred because they are smaller and less charged. A fluorescent tag is sufficiently large enough to obstruct the active part of a protein, but amino acid linkers can be used to ameliorate this effect.

Recent advances in protein engineering have generated different types of GFP proteins (Fig. 7.16). Photoactivatable GFP (PA-GFP) can switch from a dark state to fluorescent. PA-GFP was the first of these types to be used in live imaging and is useful for tracking the movement of a selected population of proteins. Photoconvertible GFP (PC-GFP) and Photoswitchable GFP (PS-GFP) proteins have also been made (Fig. 7.16). These will switch from one emission to another following activation by light or a chemical. Kaede, Eos and Dendra are examples of proteins that switch from green to red emission following activation using blue (405 nm) light. The difference between the two subtypes is Photoconvertible proteins are irreversibly converted to a red fluorescent protein (RFP) species. Such proteins can be useful for studying the activity of a subpopulation of a specific protein, for instance in trafficking or fate studies. Photoconvertible proteins are a useful addition to the live imaging toolkit because, unlike Photoactivatable proteins, they are easy to find. Photoswitchable proteins shuttle between the GFP and RFP states, these are

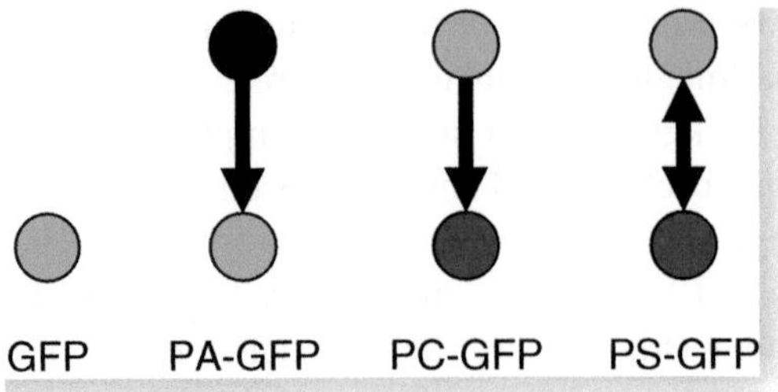

FIGURE 7.16 Fluorescent Protein Types

FIGURE 7.17 Neurons in Brainbow Mice http://suzs.tumblr.com/post/4416556844/ryan-sciandra-brainbow-is-a-term-used-to

used primarily for super-resolution imaging where blinking of a fluorophore can be used to improve localization precision.

A combination of fluorescent protein technology with gene recombination technologies (such as Cre-Lox) has lead to combinatorial expression of spectrally distinct fluorescent proteins (RFP, yellow fluorescent protein (YFP) and cyan fluorescent protein (CFP)) in neighbouring cells of model organisms, creating a 'Brainbow' or 'Confetti' of colours. These powerful model organisms have been used for live imaging to trace biological lineages or neuronal connectivity, respectively (Fig. 7.17). For more information, see http://cbs.fas.harvard.edu/science/connectome-project/brainbow.

Setting up a live imaging experiment for quantification: When setting up a live imaging experiment, it is very important to include controls as the confocal microscope set-up can vary from day to day. Lasers and detectors can fail as well as a myriad of other problems. Every live experiment should contain an internal control as a reference point. This might be the cells/tissue behaving in normal conditions. Making a cell line, tissue or organism which stably expresses fluorescent proteins tagged to the gene of interest can be beneficial for this. When quantifying experiments, it is the best practice to compare the different conditions to the reference image. The raw data from the confocal is influenced by laser power, detector sensitivity, accuracy and effective light transmission through the system. These vary week to week, and it is important to collect a set of reference images using a known sample (e.g. fluorescent beads) to keep track of this. Procedures for checking confocal microscopes are discussed in greater depth in other articles, for example ConfocalCheck (http://journals.plos.org/plosone/article?id=10.1371/journal.pone.0079879) and MetroloJ (http://imagejdocu.tudor.lu/lib/exe/fetch.php?media=plugin:analysis:metroloj:matthews_cordelieres_-_imagej_user_developer_conference_-_2010.pdf).

Fast Live Imaging

Many live applications need microscope imaging speeds on the millisecond timescale to acquire data about the intricate dynamics of biological processes. In some cases, the compromises which need to be made with point scanning simply are not compatible with the biological experiment, in that they acquire images by scanning a laser point across an image using a galvanometer mirrors, which only allow data acquisition at a rate of several microseconds per pixel. To begin to study live processes which occur faster than this, the confocal microscope needs to be redeveloped to acquire data faster. There are several different implementations of this which are detailed in Table 7.3. The major point of the table is that there is no ideal way of acquiring fast images. Resonant or slit scanning works well regardless of the specimens size, but it is still limited either by speed or by artefacts in the output image. Spinning disc confocal imaging can allow excellent quality, high-magnification images to be acquired, but in commercial implementations it may be limited to a few objectives because of the fixed pinhole size (Fig. 7.18a). Slit scanners can use slits of different shapes and sizes to scan images (Fig. 7.18b), and these 'swept field' techniques generate high-quality, multicolour fast images. Finally, light sheet microscopy (Fig. 7.18c) is a relatively new modality which instead of using an array of points of light or slits of light to simultaneously image a field of view omitting out-of-focus light, it uses a thin sheet of light. The sheet is generated by placing a cylindrical lens in front of the laser light source, meaning that only the thin section in the light sheet is illuminated, thus reducing the photodamage and stress caused by illuminating living samples with laser light. The sheet can be imaged by using a 'collection' lens that needs to be at a 90° or greater angle to the sheet. The 3D images of the sample can be collected by either stepping the sample through the light sheet or moving the light sheet and sampling lens with reference to the sample using a stepper motor. Since the entire sample is illuminated by the excitation objective, this reduces the time to scan a volume considerably. The light sheet can only be thinned to a certain degree, although use of spatial light modulators to shape the profile of the light sheet can help with this. Because of the required geometry of the excitation and detection objectives, only certain lenses can easily fit together. At present, the highest magnification which can be achieved by light sheet is 40×, generating a resolution of 1 μm. This means that light sheet microscopy gives intermediate magnification and excellent resolution and so is well suited to certain applications.

TABLE 7.3 Types of Fast Scanning Confocal Microscopes

Confocal Type	Method	Speed (Frames per Second)	Advantages	Disadvantages
Point scanner	Galvanometric mirrors scan the laser point by point across an image	5	High-precision images with no artefacts	Slow speeds. Can cause a time differential across the image
Resonant scanner	A set of resonating mirrors	25	Precision of point scanning and faster acquisition Can collect five channels simultaneously	Resonant scanner can be insensitive to light Vibration artefacts in the image can be present
Spinning disc	Excitation and emission spectra are passed through a spinning disc, which has an array of holes. This means the whole field of view is instantaneously illuminated by an array of confocal light	2000	Speed is limited by the camera acquisition Small samples, e.g. subcellular epitopes (1 μm and smaller), can be quickly imaged	The pinhole on the disk has to match the objective, so only some objectives can be used, generally 60–100× Pinhole cross-talk in disc, although dual-scanning disc and beam focusing mitigate this Disc is slightly light inefficient
Slit scanner	A thin wedge of light is formed using a cylindrical lens that converges on the focal plane at an angle determined by the numerical aperture of the objective and is swept across the imaging field	100	The 'pinholes' can be adjusted so any magnification image can be acquired Very fast data acquisition	Cross-talk between different lines The generation of the slit of light causes the point spread function to be extended in the axial direction
Light sheet confocal	A sheet of light is generated	500	Speed is limited by the camera acquisition Large samples 0.5 mm–1 μm can be imaged quickly	Data sets are very large Samples need to be dipped into water Magnification Is currently limited to 40×, so limitation in resolution

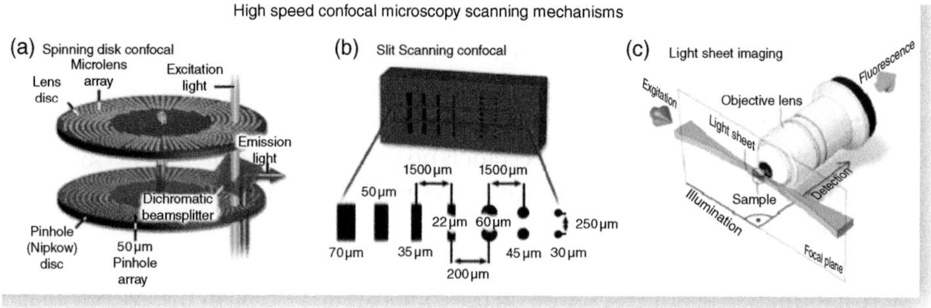

FIGURE 7.18 Schematic Diagrams of Different Fast Confocals: (a) Spinning Disk (b) Slit Scanning (c) Light Sheet

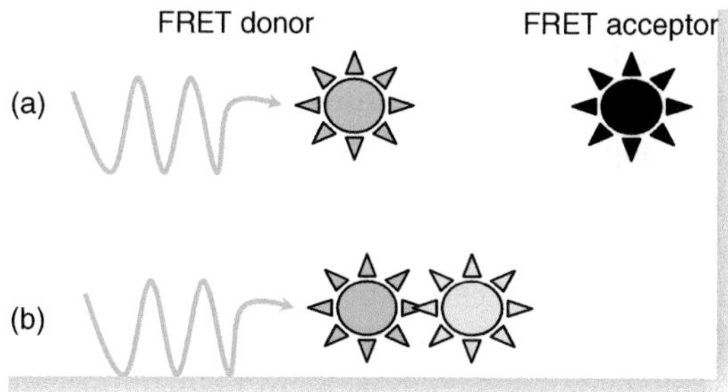

FIGURE 7.19 FRET Theory. (a) Donor Dye Is Too Far to Pass on the Energy for the Laser Light Wave onto the Acceptor. (b) Donor and Acceptor Are Close Enough Together that the Energy from the Donor Can Be Passed to the Acceptor Allowing the Acceptor's Fluorescence to Be Excited by the Energy Emitted by the Donor Dye

FRET / FRAP: Studying Interaction and Kinetics Using Confocal

Fluorescent or Forster Resonance Energy Transfer (FRET) can be used to build up detailed interaction maps of biological structures. In brief, if two fluorescent dyes with overlapping spectra come into close contact with one another (10 Å or less), the energy from the emission of dye A (donor) can be passed to dye B (acceptor) and excite fluorescence (Fig. 7.19). The emission spectrum of the donor needs to overlap with the excitation spectrum of the acceptor to allow FRET to occur.

Confocal microscopy can be very useful for FRET experiments for several reasons.

1. Confocal microscopes only use a limited number of laser lines to excite fluorescence, this limits errors in FRET due to inappropriate excitation of the acceptor dye.
2. Detection of emission light can be carefully tuned on the confocal so that the signal picked up from the acceptor dye emission can be refined.

3. Lightpaths can be easily programmed into confocal microscopes, so a setting which reads out emission from the acceptor dye when it is excited by the laser line for the donor can be easily configured.
4. Acceptor photobleaching is an easy-to-interpret form of FRET. Here, the acceptor fluorophore is bleached and (since the acceptor is no longer taking emission energy from the donor) the emission from the donor dye will go up. By using sensitive confocal detectors, in 16-bit mode, this can easily be read out.

More information about Fret set-up on the confocal can be read in the live cell imaging manual by Goldman et al. [3].

Fluorescence Recovery after Photobleaching (FRAP) and Fluorescence Photoactivation (PA)

Both FRAP and PA are used to study the mobility of proteins. Essentially, in FRAP, a small proportion of the fluorescently labelled structure is bleached away using a laser at higher power (Fig. 7.20a). In PA, a photoconvertible fluorophore is switched using an activation laser (see earlier). For these experiments to work well, the laser power for bleaching/conversion needs to be optimized. Ideally, a 50–75% bleach should occur (Fig. 7.20b). Furthermore, the bleaching/conversion can sometimes cause photodamage to surrounding areas, and so a photodamage curve should be collected as best practice to correct for this when analysing data. Following correction by quantifying fluorescent intensity changes in the region bleached/converted, it is possible to determine the stability of the protein in this region, which is the percentage of protein that recovers compared to the initial intensity and the recovery rate which is the time for half of the intensity to recover.

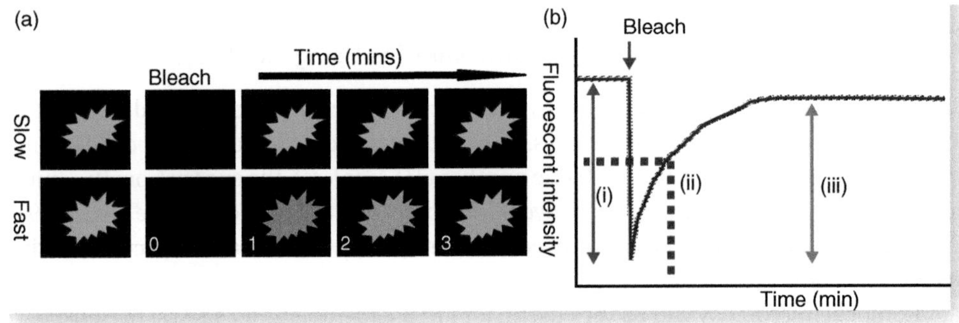

FIGURE 7.20 Fluorescent Recovery after Photobleaching. (a) Schematic Diagram of a FRAP Experiment with Slow and Fast Recovery Times. (b) FRAP Quantification (i) Shows Bleach Efficiency and (ii) Indicates the Time for Half of the Fluorescence to Recover. (iii) Once Enough Time Has Passed that the Fluorescence Has Stabilized, the Level of Fluorescence Remaining Divided by the Initial Fluorescence Intensity Shows What Percentage of the Protein of Interest Will Turn Over. This Is Termed the Mobile Fraction

More information about FRAP and PA set can be read in Goldman et al. [3] Live Cell Imaging.

Confocal Image processing

Once the confocal image has been taken, there are many next steps which follow. Some confocal images are taken for the purposes of presentation and others for quantification. When processing images, the original raw data must be kept and any adjustments should be saved in a separate file if the data are for publication. It is important to note down any manipulations that are made to the data and to explicitly state what has been done to the data when making figures for presentation in scientific publications. Many journals have strict rules about image manipulation as it is possible to alter the data so much as to falsify results (see references). This is to be avoided and it is advisable to look at the regulations for image processing that a journal has prior to making the figures for publication, just in case the method of image processing is not considered to be acceptable by that particular journal.

There are many applications that can be used to process confocal data. Generally, the manufacturer of the confocal microscope will have an offline image processing package for the data generated. It is advisable to save any data in a non-proprietary format (e.g. .tif), as confocal manufacturers can change hands, go out of business or – most likely – upgrade their equipment and image processing. Sometimes, scientific data may not be published for several years, and it is possible that a given confocal manufacturer may upgrade the equipment during that time. This could potentially mean that a file type accepted by their software at the start of a research project may not be accepted when the data are published. This can be very problematic. Saving data in a non-proprietary format that does not change over time means that the raw data can always be accessed. It also means that third-party image processing programs such as Adobe Photoshop or Illustrator can be used for making figures. This is particularly useful if data from several different pieces of equipment are combined in one scientific figure.

In addition to the manufacturer's own image analysis software, data can be processed using a range of other programs. A non-exhaustive list is shown in Table 7.4.

Image Processing for Presentation

The confocal image comprises an array of pixels. Each pixel has a given intensity value which corresponds to the amount of emitted light detected in that specific space. The intensity value is numeric, so the image that a confocal produces can be considered to be a matrix of numbers. In a nutshell, image processing is carrying out mathematical operations on the matrix of numbers which makes up a scientific image.

For presentation purposes, the number of pixels can be reduced to compress data, or part of the image which is particularly of interest can be cropped out. It is not acceptable to increase the number of pixels in an image, as this would mean that new data are being made up. If a zoomed-in image is required, then it is preferable to acquire it on the confocal. However, increasing (binning) the size of the pixels (i.e. from 80 to 150 nm) but not the

TABLE 7.4 Image Processing Packages for Confocal Microscopy

Image Analysis Program	Function	Website
ImageJ/Fiji	Free open-source scientific image analysis software	http://imagej.nih.gov/ij/http://fiji.sc/Fiji
Ilastik	Interactive Learning and Segmentation Toolkit	http://www.ilastik.org/
Image-Pro Plus	Count, measure and classify objects; and automate work	http://www.mediacy.com/index.aspx?page=IPP
SlideBook	Image processing and automated data analysis	https://www.intelligent-imaging.com/slidebook.php
Metamorph	Image processing and automated data analysis	http://www.moleculardevices.com/products/software/meta-imaging-series/metamorph.html
Huygens	Deconvolution of microscopic images	http://www.svi.nl/HuygensEssential
Imaris	Data visualization, analysis, segmentation and interpretation of 3D/4D microscopy data sets	http://www.bitplane.com/imaris/imaris
Volocity	Data visualization, analysis, segmentation and interpretation of 3D/4D microscopy data sets	http://www.perkinelmer.co.uk/pages/020/cellularimaging/products/volocity.xhtml

number of them is acceptable, but be aware that there may be a loss of resolution in the image.

It is acceptable to contrast stretch an image. This is done by adjusting the brightness and contrast. Alternatively, the range of intensities which are displayed on the screen can be adjusted. Since an image will comprise of a range of intensities from black (0) to very bright (which depends on the bit depth), it is possible to remove background from the image by simply setting the blackest point displayed as a value higher than 0. For images which look too dim, the brightest point can be reduced from the maximum value (255 for 8-bit images) to a lower value. This is not manipulating the data as the raw data remain unchanged. Adjusting the displayed contrast simply shows what is present in the data more clearly. As most fluorescent images are taken in the dark, frequently there are issues with the image looking more or less bright in daylight because the eye has dark adapted when the image is taken.

The gamma function in an image shows the mathematical relationship between the intensity gradations. The gamma value should be left as 1 for fluorescent images as it shows a 1:1 relationship between the gradations in the image between black and white. Altering the gamma value falsely enhances certain features by non-linear stretching of the intensities (see [4]).

Quantification of Confocal Images

When quantifying image data, it is important that identical manipulations are carried out on the whole of the data set, much in the same way as the sample preparation and data acquisition on the confocal are kept the same.

It is very common to use image maths to remove background from the image. This can be done by subtracting a constant value from all of the pixels which make up an image. More complex background subtraction, to get rid of graininess in images or uneven illumination, can be carried out using rolling ball filters and median filters or by taking an image of an unstained area and subtracting this from the image to be processed. Images can also be filtered to improve the identification of features. In general, a sharpen or high-pass filter will increase the sharpness of an image. Low-pass or blur features will make the image look softer and iron out any unevenness. 'Top-hat' is a very useful filter, as it removes the speckle from an image without smoothing the edges of larger objects. This filter is also very useful for removing PMT noise that has not been removed by frame averaging. All image filters consist of a kernel, generally a 3×3 array of numbers, from which the image matrix is processed. If possible, it is advisable to determine what the image processing kernel is and to record this in the methodology section of any scientific publications subsequently written.

Deconvolution of Confocal Images

Although confocal images will have almost no out-of-focus light present, it is still possible for there to be artefacts due to neighbouring points. Wear and tear on an objective lens means that over time it will not produce the precise PSF that it could at the time of manufacture. It is important to routinely perform quality control checks on the confocal to ensure that it is properly aligned and the PSF has not deviated too much from the theoretical one for the system. This can be done by using multicolour beads such as TetraSpeck™ beads. The real PSF of the system can be determined by taking a z stack through the bead. The size of the bead ought to be equivalent or slightly smaller to the minimum resolvable size of an object by a lens. By using the PSF image generated, it is then possible to correct any artefacts in the data by deconvolving the output image against the PSF. This removes any residual blur or out-of-focus light. There are a number of packages which can deconvolve data, some of which are listed in Table 7.4 (image processing packages). Deconvolution is a mathematically rigorous process that involves transforming the image and the PSF into Fourier space and then convolving the two images together to restore any lost light and remove any out-of-focus light. A computer which is equipped with a fast processor (Core i3 or more) and a large amount of RAM (8 GB or more) is required for such operations. In general, the manufacturers of deconvolution packages will provide advice on the computer requirements.

Presentation of Confocal Images in 3D

Three-dimensional rendering of confocal data requires considerable computational processing power. Similar to deconvolution, it is necessary to have a fast and powerful computer to carry out these analysis. Rendering data in 3D and controlling the light path, image processing and segmentation are more mathematically rigorous than 2D data; so for those without in-depth knowledge of computer programming, it is recommended that a professional package is used for this. These are listed in Table 7.4 (image processing packages). Using these image processing tools, it is possible to acquire high-quality images which very clearly show the 3D structures of tissue. These packages also allow the distance between and volume of the structures to be measured. Combining *z* stacks, live cell imaging and 3D quantification using intra-vital microscopy are all key areas advancing today's research, giving in-depth powerful information about the physiological processes of life.

REFERENCES

1. Spector, D.L. and Goldman, R.D. (2006) *Basic Methods in Microscopy: Protocols and Concepts from Cells: A Laboratory Manual*, Cold Spring Harbor Laboratory.
2. Eliceiri, K.W. (2004) Molecular expressions: Exploring the world of optics and microscopy. http://micro.magnet.fsu.edu/. *Biology of the Cell*, **96**, 403–405.
3. Goldman, R.D., Swedlow, J.R. and Spector, D.L. (eds) (2010) *Live Cell Imaging: A Laboratory Manual*, Second edn, Cold Spring Harbor Laboratory.
4. Rossner, M. and Yamada, K.M. (2004) What's in a picture? The temptation of image manipulation. *The Journal of Cell Biology*, **166** (1), 11–5.

FURTHER READINGS

Alison, J. (2006) Seeing is believing? A beginners' guide to practical pitfalls in image acquisition. *The Journal of Cell Biology*, **172** (1), 9–18.

Allan, V.J. (1999) Basic immunofluorescence, in *Protein Localization by Fluorescence Microscopy—A Practical Approach* (ed. V.J. Allan), Oxford University Press, Oxford, UK, pp. 1–26.

Murray, J.M., Appleton, P.L., Swedlow, J.R. and Waters, J.C. (2007) Evaluating performance in three-dimensional fluorescence microscopy. *Journal of Microscopy*, **228** (Pt 3), 390–405.

Pawley, J. (ed.) (2006) *Handbook of Biological Confocal Microscopy*, Springer.

Rossner, M. and O'Donnell, R. (2004) The JCB will let your data shine in RGB. *The Journal of Cell Biology*, **164**, 11.

Swedlow, J. et al. (2003) 300 (5616): 100–102 Science Magazine (Apr 4).

CHAPTER EIGHT

8

Ultrastructural Immunochemistry

Jeremy Skepper and Janet Powell

Cambridge Advanced Imaging Centre, Department of Anatomy, University of Cambridge, Cambridge, UK

INTRODUCTION

The landmark publication by Coons et al. in 1941 [1] demonstrated that an antibody conjugated to a fluorochrome retained its ability to recognize and bind tightly to its antigen. This was arguably the key event in the development of immunofluorescence microscopy. Once the general principles for immunochemical staining were established, the introduction of particulate markers for transmission electron microscopy (TEM) followed quickly. Ferritin, a moderately electron-dense protein with a 4-nm iron core, was the first to be chemically conjugated to antibodies by cross-linkage between its protein shell and an antibody some 18 years later [2, 3].

The use of colloidal gold technology was undoubtedly the most significant event in the development of immunochemistry. It was demonstrated that protein molecules, including antibodies, could be adsorbed on to the surface of gold particles with little or no loss of their

Immunohistochemistry and Immunocytochemistry: Essential Methods, Second Edition. Edited by Simon Renshaw.

biological activity [4]. Gold particles are particularly useful for TEM studies, as they scatter electrons strongly and even small particles are clearly visible under the electron microscope. Next, the ability to produce colloidal gold particles with different mean sizes and non-overlapping size–frequency distributions [5] brought the potential for immunochemical staining of multiple antigens on the same thin section, a development that significantly improved the value of the method. The application of random sampling strategies and the use of unbiased stereology have allowed us to make quantitative comparisons of labelling density [6, 7]. In some instances, it is possible to estimate the concentration of the antigen within its host tissue by making a comparison of labelling density on the specimen with that over an internal standard containing a known concentration of antigen [8].

Before proceeding to immunogold staining, it is important to gather as much information as possible about your antibody and its respective antigen. Where is it likely to be located? Is the antigen extracellular, intracellular, membrane-associated or a soluble component of the cytoplasm? Is it there in significant quantities? Is it sequestered at high concentration in any specific subcellular compartment, such as the mitochondria or the nucleus? How vulnerable to fixation and embedding is the antigen of interest? Information on the specificity of antibodies from Western blotting is valuable, but even if antibodies work well in Western blotting, there is no guarantee that they will be useful for immunochemistry. Antibodies that 'work well' on blots frequently have to be used at concentrations of up to three or more orders of magnitude greater for immunofluorescence and even more for immunogold staining studies. Some antibodies simply cannot be used for immunochemistry! The degree of resistance of the antigen to fixation is also a key issue. In general, the stronger the fixative that can be used, the better the ultrastructure of the tissues and in particular that of membranes. Unfortunately, the opposite generally applies to the ability of the antibody to bind its antigen. This also relates to cryotechniques. It is possible to freeze and embed tissue at low temperature without any chemical fixation, but ultrastructure is always compromised to some degree. If a chemical fixative is added to the substitution mixture, the frozen tissue is dehydrated and fixed at the same time. Fixation is also less efficient at low temperature. The gold standard is to find the appropriate compromise that allows one to answer the biological question.

Fixation and Its Effect on Antigen–Antibody Binding

Glutaraldehyde and formaldehyde are the two fixatives in most common use, either individually or in combination. Formaldehyde is a monoaldehyde that interacts principally with proteins forming methylene bridges or polyoxymethylene bridges in a concentration-dependent manner. Glutaraldehyde is a dialdehyde that gives superior ultrastructural preservation but can cause significant conformational changes to the tertiary structure of proteins. This frequently compromises the ability of an antibody to bind to its antigen. For a detailed discussion on the chemistry of fixation, see [6, 9] and [10] and p 37. In this context, a 'stronger' fixative will be regarded as a fixative containing higher concentrations of the reactive aldehydes.

The ability of the antibody to bind its antigen may be lost at several key stages of processing for TEM: during chemical fixation, dehydration in organic solvents, infiltration with epoxy or acrylic resin, or during heat curing or polymerization of the resin. New antibodies should always be tested by a method that does not amplify signal, such as a species-specific, fluorescent secondary antibody method, as there are only minor options for signal amplification in electron microscopy. There are several key questions to ask if an antibody has been used for prior immunochemical studies:

- Does the antibody work only on unfixed or cold acetone/alcohol-fixed cryostat sections or cell cultures? If the answer is yes, this antibody may only be usable in methods employing cryoimmobilization and freeze substitution in pure organic solvent rather than after chemical fixation.
- At what strength and duration of fixation will the antigens survive and still bind with specificity to their respective antibodies?
- Does the antibody work on sections of formalin-fixed, paraffin wax-embedded tissue, without antigen retrieval treatment?

It is wise to undertake a systematic evaluation of fixation on a tissue known to contain significant amounts of the antigen under study. This constitutes a positive control, which is highly desirable (if not essential) in any rigorous study and may well provide critical information. Fixation of tissues and organs is best carried out by vascular perfusion [9, 10]. This minimizes the diffusion distance into the tissue for the fixative, thus avoiding a gradient of fixation and also a potential gradient of immunoreactivity. There are, however, circumstances where fixation by perfusion is impossible or may be undesirable. Bendayan et al. [11] showed that immunogold staining of serum albumin in glomerular capillaries was reduced dramatically after perfusion fixation, presumably because the serum albumin molecules were washed out during exsanguination.

If it is not possible to fix by perfusion, for example when working with human tissues from surgery or biopsies, samples should be small. A simple method of achieving uniformity of fixation is to glue two safety razor blades together at the shank to produce two parallel blades <1 mm apart. Tissues are sampled using a gentle slicing motion, rather than by applying significant vertical force, in order to minimize mechanical damage. Alternatively, a tissue chopper or vibrating microtome can be used to cut thin slices. The danger with the latter is the tissues will have a longer post-mortem delay before fixation so slicing should take place in an oxygenated medium wherever possible and not in a simple PBS or normal saline. Cells in culture are much easier to deal with, as diffusion distances for fixatives are minimal. They should be cooled to 4 °C and rinsed in normal saline (0.9%, w/v, sodium chloride) before fixation. Non-adherent cells can be fixed in suspension, while adherent cells should be fixed *in situ* for 30–120 min and then scraped free from their substrate.

Both the temperature and duration of fixation should be standardized to maintain uniformity between experiments. We carry out initial fixation tests at 4 °C for no more than 120 min for tissues and 30–120 min for cell cultures, while others prefer fixation at

37 °C [12], but only trial and error will determine the appropriate compromise between structural preservation and the ability of the fixed antigen to bind antibody. Safety is a major issue when fixation is carried out at temperatures above 4 °C as aldehydes are volatile. Formaldehyde is a known carcinogen and glutaraldehyde can cause occupational asthma. If fixation is performed at an elevated temperature, it must be carried out in a fume hood.

It is convenient to test new antibodies on adherent cell cultures expressing the antigen of interest grown on glass coverslips or on cryostat sections. Cells are grown to near-confluence on 19-mm-diameter coverslips of No. 1 thickness and fixed for 30–120 min at 4 °C. Naturally, if the antigens will survive longer periods of fixation (up to 4 h), then ultrastructural preservation will be even better. They are rinsed in four to six changes of buffer before being stained immunochemically. Alternatively, fixed tissues are infused with 20% (w/v) sucrose and frozen to prepare cryostat sections. We routinely store a range of fixed and unfixed tissues (myocardium, liver, gut, placenta, etc.) under liquid nitrogen so that material is always available for testing new antibodies. An initial test is carried out comparing the effects of weak and strong fixatives with a short or long duration using the following solutions:

(a) 1% (w/v) formaldehyde in 0.1 M PIPES, HEPES or cacodylate buffer (pH 7.4) containing 3 mM/L calcium chloride.
(b) 4% (w/v) formaldehyde in 0.05 M PIPES, HEPES or cacodylate buffer (pH 7.4) containing 3 mM/L calcium chloride.
(c) 8% (w/v) formaldehyde in 0.05 M PIPES, HEPES or cacodylate buffer (pH 7.4) containing 3 mM/L calcium chloride.
(d) 3% (w/v) formaldehyde plus 0.05–0.1% (w/v) glutaraldehyde in 0.05 M PIPES, HEPES or cacodylate buffer (pH 7.4) containing 3 mM/L calcium chloride.

The above-mentioned fixatives are listed in an ascending order of potential ultrastructural preservation, but probably in a descending order of antibody binding. Tissue sections or cultured cells fixed in solutions (a), (b) and (c) are ready for immunochemical staining after rinsing in buffer. Those fixed in solution (d) must be incubated in 0.5% (w/v) sodium borohydride for 5–10 min and rinsed in buffer to quench the autofluorescence generated by glutaraldehyde. Test parameters should also include a range of dilutions of primary antibodies, usually 1:5, 1:25, 1:100 and 1:1000 for monoclonal antibodies and 1:50, 1:250, 1:1000 and 1:5000 for polyclonal antibodies.

Antigens that survive very strong fixation and embedding in paraffin wax may well survive ambient temperature dehydration and embedding in thermally cured epoxy resin after secondary fixation with osmium tetroxide. In this method, the osmium tetroxide is removed from the superficial layers of the section by treatment with periodic acid and/or sodium metaperiodate [13]. Thin sections are floated on drops of the oxidizing agent of choice, for example 5% (w/v) sodium metaperiodate, for 10–20 min and then rinsed thoroughly with ultrapure water before commencing immunochemical staining. Periodic acid and sodium metaperiodate are both oxidizing agents with differing efficacies. Some workers just use sodium metaperiodate while others suggest that a sequential treatment with both produces stronger immunochemical staining. We tend to use a single treatment with sodium metaperiodate, which removes osmium tetroxide from the surface of the thin section, and in some cases this will enhance the binding of an antibody to its antigen at that surface. Antigens that withstand modest fixation but not paraffin wax embedding are generally more suitable for embedding in acrylic resin at ambient or at low temperature. Antibodies that only work on unfixed cryostat sections may work in cells or tissues that have been cryoimmobilized, dehydrated by freeze substitution and embedded at low temperature. However, there is no guarantee that the integrity of the antigen will not be compromised by the subsequent dehydration, embedding and curing, or polymerization of the resin.

It may also be necessary to use a stronger fixative to retain antigens that are freely soluble in the cytoplasm [14]. It is interesting to note that cells with a high content of secretory granules and endoplasmic reticulum often show reasonable preservation, even after weak fixation, particularly if they are processed subsequently using the freeze substitution and low-temperature embedding route. This may be at least partly due to their high protein content (see Fig. 8.1). As the strength of fixation is reduced, ultrastructural

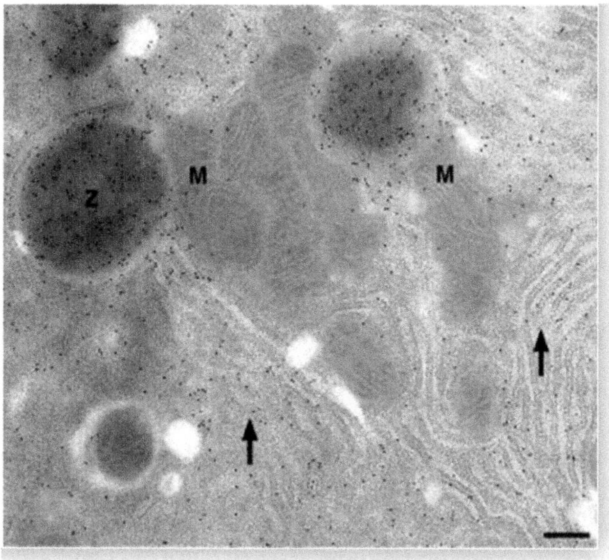

FIGURE 8.1 Thin Section of a Rat Pancreatic Acinar Cell. The Section Was Fixed in 3% Formaldehyde, Cryoprotected in 30% Polypropylene Glycol, Dehydrated by Freeze Substitution and Low Temperature Embedded in Lowicryl HM20. Cells Were Immunostained for the Presence of Amylase. Gold Particles Indicate the Rough Endoplasmic Reticulum (Arrows) and Zymogen Granules (Z). Mitochondria (M) Are Unlabelled, Showing That Non-Specific Labelling Is Low. Bar, 200 nm

preservation becomes poorer, particularly that of membranes. The low-temperature methods compensate to some degree, but it is inevitable that weaker fixation means poorer preservation. The method that retains the best membrane preservation is undoubtedly the ultrathin, thawed cryosection or 'Tokuyasu' method [15], but again stronger fixation gives better preservation. A comprehensive description of this technique is beyond the scope of this chapter, and the reader is referred to [6] and the seminal papers by Peters et al. [12] and Liou et al. [16].

Controls

Controls are essential but ostensibly simple, requiring tissue or cells expressing significant amounts of the antigen in question as a positive control.

A negative control is equally important, as it will indicate whether there is non-specific binding of primary or secondary antibodies. Sections of cells should also be exposed routinely to the secondary antibody alone to be certain that there is no non-specific binding to any component of the tissue. In a recent unpublished study carried out with Raghu Padinjat of the Babraham Institute (Cambridge, UK), we encountered a most elegant example of a combined positive and negative control in adjacent cells of the same tissue. Ommatidia are the light-sensing structures of the Drosophila eye. Each ommatidia contains seven rhabdomeres (see Fig. 8.2), which have extensive membrane systems derived from microvilli. The membranes of six of the rhabdomeres are rich in rhodopsin, a light-absorbing pigment, while the seventh rhabdomere contains no rhodopsin (see Fig. 8.2), making it an ideal negative control. If excessive non-specific binding of the primary or secondary antibody is apparent, protein can be added to the buffers to inhibit it competitively. Various proteins are used for this purpose, but in our hands BSA or coldwater-fish gelatin, both used at 0.1–4% (w/v), give the most consistent results.

Remember that this will also competitively inhibit specific binding, so the concentration of blocking protein should be kept as low as possible.

Why Do We Need to Use Electron Microscopy?

The answer to this is resolution. In the light, confocal and two-photon microscopes, resolution is diffraction limited to 180–200 nm in the x–y axes, dependent on the numerical aperture of the objective lens and the wavelength of light used to generate the image. Resolution in the z-axis is much poorer at 500–600 nm. However, there are techniques that can bypass these limitations. These include total internal reflectance (TIRF) microscopy, stimulated emission depletion microscopy and 4Pi microscopy. TIRF [17] and 4Pi [18] microscopy can exceed 100 nm resolution, but with severe limitations on specimen and lens geometry in 4Pi microscopy and in the depth of imaging into a sample with TIRF microscopy. Stimulated emission depletion microscopy [19] can exceed 50 nm resolution but requires a very high signal-to-noise ratio and an almost ideal sample.

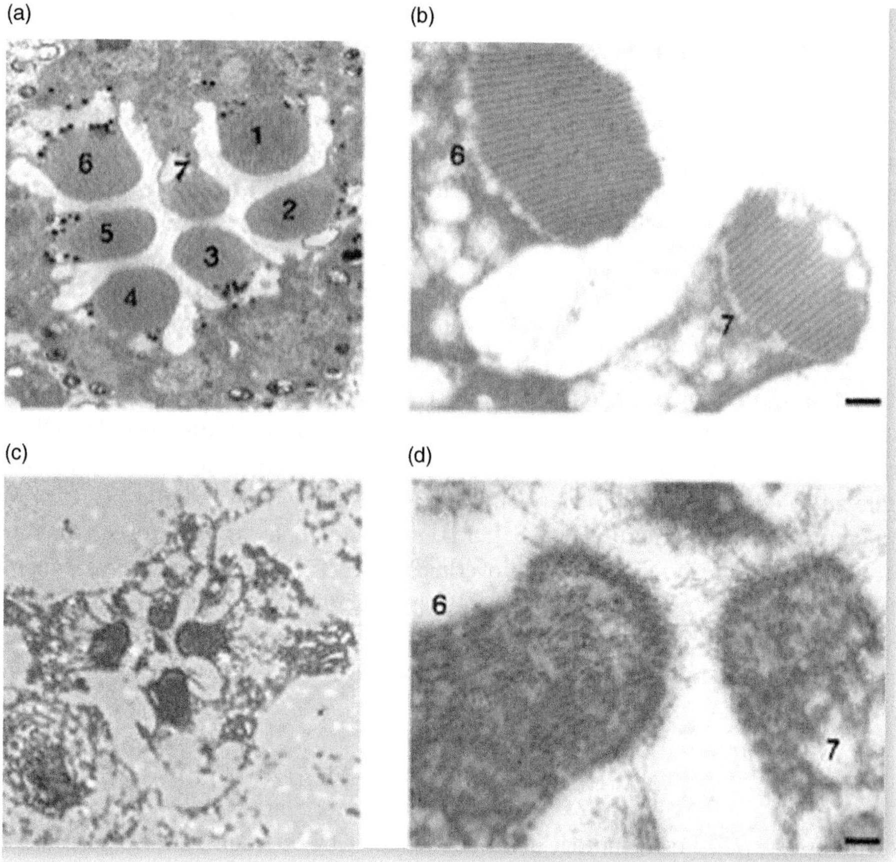

FIGURE 8.2 Thin Sections through Single Ommatidia from a Wild-Type or Mutant Drosophila Eye, Immunolabelled for Rhodopsin. The Eyes Were Fixed in 3% Glutaraldehyde, Osmicated and Embedded in Spurr's Resin (a, b; Wild Type) or Fixed in 4% Formaldehyde and Embedded in LR White (c, d; Mutant). Each Ommatidia Contains Seven Rhabdomeres (a). Rhabdomeres 1–6 Express Rhodopsin, while Rhabdomere 7 Does Not (b). In the Mutant Eye, Ommatidia Are Deleted or Altered (c). The Rhabdomeres Are Also Structurally Altered, but Their Staining Pattern for Rhodopsin Remains Unchanged, with No Expression of Rhodopsin in Rhabdomere 7 (d). Bars, 200nm

Quantification

If a single compartment is being stained immunochemically and the biological question is simply whether or not there is staining over that compartment, then quantification is unnecessary. If label (staining) density is low and you wish to make a comparison between multiple compartments in control and experimental subjects, then quantification is essential. Quantification of label density is simple and strengthens data immensely. It is a simple extension of stereology, which is used to gain three-dimensional data from what are effectively two-dimensional sections. When comparing mutant and wild-type organs, it is desirable to start the comparison with an estimate of the volume or reference space of the organ itself. If the organ of the mutant is halved in volume but the percentage of it occupied by a specific cell type is doubled, the total volume of that cell type is unchanged. This phenomenon is known as the 'reference trap' [20]. A typical example might be to examine the effect of a mutation on the distribution of rhodopsin in the eye of Drosophila (see Fig. 8.2). After fixation and embedding in a suitable resin, serial sections (2 μm in thickness) are cut through the eye and the Cavalieri method [21] is used to estimate the volume of the eye in mutant and wild-type flies. At four randomly selected levels through the layer containing the rhabdomeres, thin (50–70 nm in thickness) sections are cut, immunogold labelled for rhodopsin and contrast counterstained with uranyl acetate and lead citrate. Both uranyl acetate and lead citrate impart contrast to the tissue. They are viewed at 80–120 kV in a transmission electron microscope using a 10- or 20-μm objective aperture to maximize contrast. A quadratic (square) lattice is overlaid on the TEM image and the volume fraction (Vv, expressed as a percentage) of the eye occupied by ommatidia and rhabdomeres is estimated by point counting [21], that is counting the number of points (P) from the intersections of the counting lattice that overlie the area of interest (i). The formula for this calculation is as follows:

$$Vv_{\text{rhabdomere}}(\%) = (Pi_{\text{rhabdomere}}/Pi_{\text{total}}) \times 100$$

where $Vv_{\text{rhabdomere}}$ is the percentage of the eye occupied by rhabdomeres, $Pi_{\text{rhabdomere}}$ is the number of lattice intersections overlying rhabdomeres and Pi_{total} is the total number of lattice intersections overlying all compartments of the eye. Therefore, if $Pi_{\text{rhabdomere}} = 10$ and $Pi_{\text{total}} = 100$, 10% of the eye is occupied by rhabdomeres.

The light-absorbing pigment rhodopsin is associated with the photoreceptor membranes of the rhabdomere, and immunogold label density can be calculated as the number of gold particles per unit area (number/μm^2) of rhabdomere. Number/unit area can be estimated by randomly selecting squares of the counting lattice overlying the areas of interest (rhabdomeres) and counting all of the gold particles within the square and those within the frame that also intersect with two of the four boundary lines of the counting frame. This is known as the forbidden line rule [22] and prevents the underestimation of particle density that occurs if particles are only counted and if they are within the square but not touching the counting frame. Similarly, the number will be overestimated if all particles, including those intersecting all four boundaries, are included. The area of an individual square counting frame of a quadratic lattice is $D2$, where D is the distance between two intersections of the lattice. Therefore, gold label density can be estimated by summing the number of gold particles in, for example, 10 randomly selected test frames and dividing that by the total area of those frames in microns.

The procedure described earlier gives a parametric estimate of gold labelling density over a structure or series of subcellular compartments. Mayhew and co-workers [23, 24] have suggested a non-parametric alternative that estimates the 'relative labelling density' between compartments and between control and experimental subjects.

METHODS AND APPROACHES

As one would expect with a technology that is more than 40 years old, the number of methods and their variants is extensive. Many methods are incompletely described, particularly those in research publications with restrictions on space for methods, and to the novice many may appear to be a combination of 'cookery' and 'witchcraft'. This chapter will describe four methods that are in common use. For more comprehensive descriptions of the range of techniques available, see [6] and [25].

Three of the methods described here are post-embedding methods. In these methods, the cells or tissues are fixed chemically or cryoimmobilized, dehydrated and embedded in epoxy or acrylic resins. Thin sections (50–70 nm in thickness) are cut using an ultramicrotome with a diamond knife, using a water bath to collect the sections as they slide off of the knife. The sections are stretched with solvent vapour or a heat source and collected on to either bare or plastic-coated nickel grids. The sections are then stained immunochemically with primary antibodies raised against antigens exposed on the surface of the sections. The primary antibodies are then visualized by staining immunochemically with secondary antibodies raised against the species and isotype of the primary antibodies, conjugated to colloidal gold particles. The immunochemically stained sections are then contrast stained with salts of uranium (uranyl acetate) and lead (lead citrate) to reveal the ultrastructure of the cells and are finally viewed by TEM.

The fourth method is effectively a pre-embedding method that is used if antigens are damaged by resin embedding or if the best preservation of membranes is required. Cells or tissues are fixed as strongly as possible and then treated with a cryoprotectant, which is usually a mixture of sucrose and polyvinylpyrrolidone. They are frozen on to pins in liquid nitrogen and sectioned at −100 °C. The frozen sections are thaw-mounted on to Formvar/nickel film grids and the cryoprotectant is removed by floating the grids on drops of PBS. The immunogold staining is performed on the non-embedded section, and it is subsequently contrast counterstained and infiltrated with a mixture of methylcellulose and uranyl acetate. Methylcellulose reduces shrinkage of the section when it is air-dried before viewing by TEM.

Epoxy Resin Section

Chemical fixation and embedding in a highly cross-linked epoxy resin is the method of choice for optimal ultrastructure and stability of the thin section in the electron beam. Paraffin wax cannot be used for TEM as it is impossible to cut thin enough sections because the wax is too soft. Even if it were possible to cut sections that were thin enough, the wax would evaporate in the electron beam and contaminate the column of the microscope. Immunogold staining of thin epoxy resin sections is useful if the antigen of interest is very resistant to fixative or if only archived material that was fixed primarily for ultrastructural studies is available. It would be ideal if we could fix and embed tissue to produce the very best ultrastructure, yet leave the tissue with sufficient antigenicity for it to be immunochemically stained. This would optimally include fixation in a high concentration of glutaraldehyde (2.5%, w/v, or higher, see p 40) followed by secondary fixation with osmium tetroxide and bulk staining in uranyl acetate. Osmium tetroxide fixes by binding to double bonds in unsaturated fatty acids, retaining them in the subsequent dehydration in organic solvent. It adds positive contrast because it is a heavy metal that scatters electrons. Similarly, uranyl acetate acts as both a fixative and a stain, as it helps retain phospholipids and adds contrast to the thin sections by scattering electrons. The fixed tissue is dehydrated in an organic solvent infiltrated with an epoxy resin, which is thermally cured at 60 °C for up to 48 h. Epoxy resin monomers are joined end to end to form long-chain polymers, which are in turn cross-linked to adjacent polymers during the curing process. This makes them very stable in the transmission electron microscope but hinders access of the antibody to the antigen. Some antigens do survive this treatment, notably small peptide hormones or neurotransmitter substances that are found highly concentrated in secretory vesicles (see Fig. 8.3). High concentrations of glutaraldehyde are used in protocols for immunochemical staining of amino acid neurotransmitters, such as glutamate and γ-aminobutyric acid [8]. This appears to be necessary to ensure they are not physically extracted during subsequent dehydration and embedding. It is generally necessary to remove osmium tetroxide from the superficial regions of the thin section to be immunochemically stained. This is readily achieved by treatment with one or a combination of the following oxidizing agents: 10% (v/v) hydrogen peroxide [26], 4% (w/v) sodium metaperiodate [27] or 1% (w/v) periodic acid [8]. This pre-treatment of resin sections of tissues fixed to maximize ultrastructural preservation has been used to a great effect for the study of secretory proteins, peptides and neurotransmitters [13, 28–31]. However, these antigens are generally present in very high local concentrations within secretory granules. Oxidizing agents attack the hydrophobic alkane side chains of epoxy resins, which make the sections more hydrophilic [26]. This allows more intimate contact between the immunochemical reagents and the antigens exposed at the surface of the sections.

The Acrylic Resins London Resin (LR) White and Gold

LR White was introduced as a low-toxicity alternative to epoxy resins, which frequently contained carcinogens [32]. It contains an initiator and can be polymerized by the application of heat at 48–50 °C or by chemical catalysis at temperatures as low as −15 to −20 °C, albeit exothermically. At temperatures below −15 °C, its viscosity is very high and infiltration of the resin into the tissue becomes problematic. LR Gold is less hydrophobic and can be polymerized by photoinitiation using benzoin methyl ether as

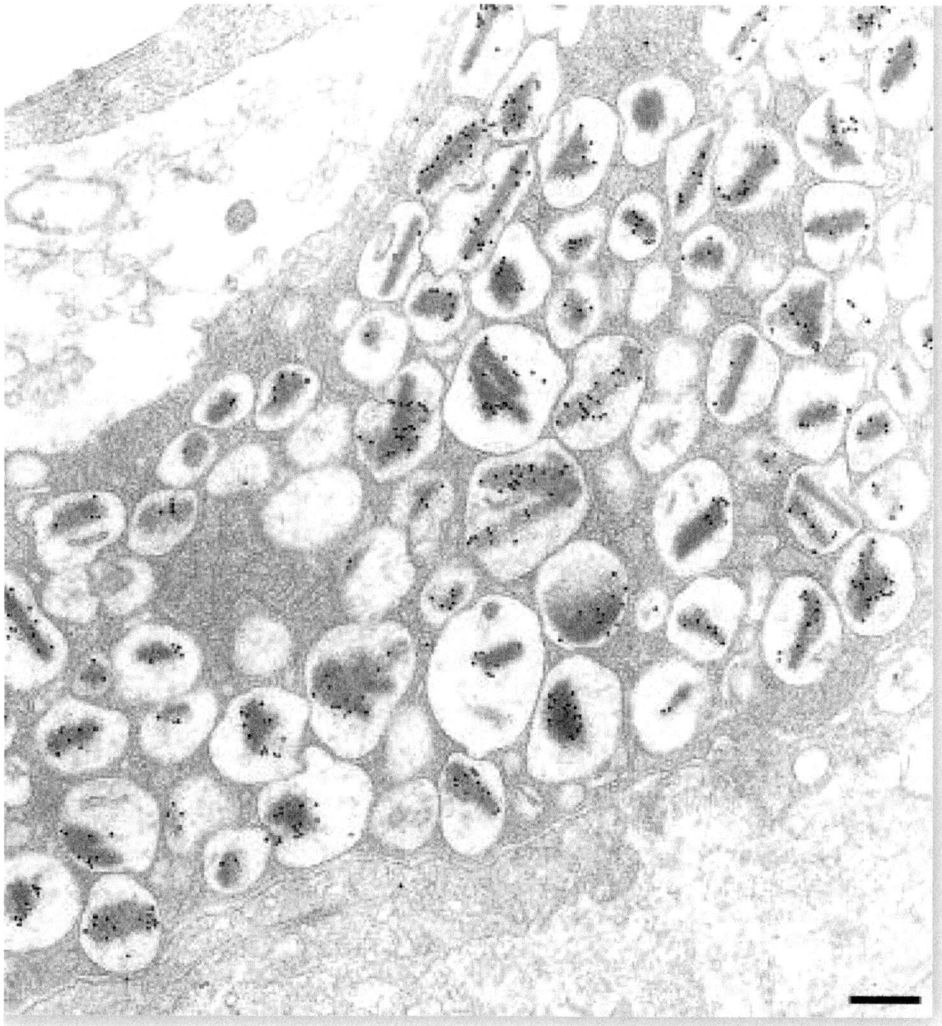

FIGURE 8.3 Thin Section through a Rat Pancreatic β-Cell. The Tissue Was Fixed in 4% Glutaraldehyde/1% Osmium Tetroxide, Bulk Stained in Uranyl Acetate and Embedded in Spurr's Resin. The Section Was Treated with Sodium Metaperiodate before Immunostained for Insulin. The Crystalline Cores of the Secretory Granules Are Heavily Labelled with Gold Particles. Bar, 250 nm

a catalyst down to −25 °C. It should be noted that at temperatures below −18 °C, the initiator can spontaneously come out of solution. Unlike the simplest acrylic resins, in which monomers are polymerized to form long chains, the LR resins contain aromatic cross-linkers to improve the stability of the sections under the electron beam. Both LR White and Gold have very low viscosity and readily penetrate even into dense tissue.

Aldehyde-fixed tissue is dehydrated and embedded in the acrylic resin without secondary fixation in osmium tetroxide. The tissue is dehydrated in ethanol, impregnated in acrylic resin and polymerized under vacuum or in a nitrogen atmosphere. The inert atmosphere is necessary because oxygen inhibits the polymerization of these resins. Acetone is not recommended as a dehydrating agent because it can act as a scavenger of free radicals, which can interfere with the polymerization of the resin. A convenient method for flat embedding is to place tissue in an aluminium weighing boat and exclude oxygen by dropping a piece of Melinex sheet or a Thermanox coverslip on to a positive meniscus of resin. Polymerization can be initiated chemically or photolytically at 4–20 °C or thermally at 48–60 °C. We find it convenient to flat embed in polyethylene or polypropylene caps such as the Kapsto GPN 600 that can be obtained from Poeppelmann. These can be half-filled with LR White and polymerized in the Leica AFS under nitrogen gas for 10 h at 45 °C followed by 10 h at 55 °C. It is claimed that reducing the temperature during polymerization enhances antigen survival. If this is the case, in most instances the gain is likely to be marginal if the difference is a drop from 60 °C to ambient temperature. Membrane preservation can be improved by bulk staining with uranyl acetate (see Fig. 8.4) before dehydration [33]. The advantages of this method are as follows: (i) the

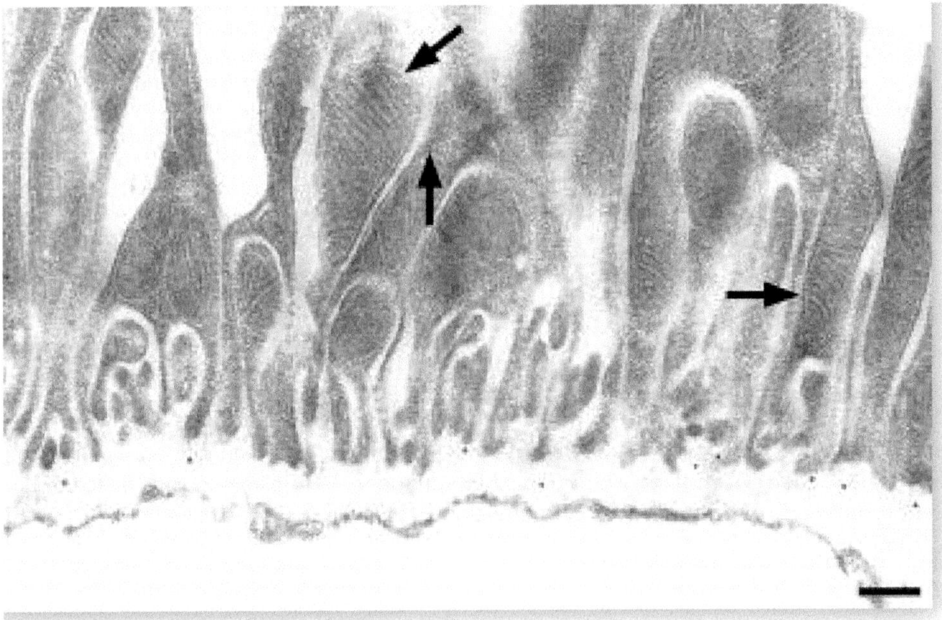

FIGURE 8.4 Thin Section through a Proximal Convoluted Tubule of a Rat Kidney. The Tissue Was Fixed by Immersion in 2% Formaldehyde and Embedded in LR White after Bulk Staining in Uranyl Acetate. The Basal Lamina Is Labelled with Gold Particles after Immunostaining for Laminin. Despite the Weak Fixation, the Outer Mitochondrial Membranes and Cristae of Mitochondria (Arrows) Can Be Clearly Distinguished. Bar, 250 nm

polymerized acrylic resin matrix is 'looser' than that produced in a cured epoxy resin; and (ii) the sectioning properties are different from those of epoxy resins and the antigens revealed at the surface of the section may be more accessible to the antibody molecules.

Freeze Substitution and Low-Temperature Embedding in Lowicryl HM20

Lightly fixed pieces of tissue are cryoprotected by immersion in 30% (v/v) glycerol or polypropylene glycol [34]. The cryoprotectant provides many nucleation sites within the tissue, so that, even when slower freezing methods are used, the small ice crystals formed are unresolvable by TEM at the magnifications used for most immunogold staining studies. The cryoprotected tissues are mounted on small pieces of aluminium foil or on pieces of Millipore filter. They are quench frozen by plunging into liquid propane cooled in liquid nitrogen. Adequate freezing can also be obtained using nitrogen slush or even liquid nitrogen for very small samples. Alternatively, monolayers of cells or thin slices can be frozen rapidly, freeze substituted and low-temperature embedded with no chemical fixation at all [35, 36] (see Fig. 8.5). This method is therefore suited to antigens that are sensitive to aldehyde fixation.

The frozen tissue samples are transferred under liquid nitrogen to vials half-filled with frozen methanol or methanol containing low concentrations (0.01–1%, w/v) of uranyl acetate. Chilled metal forceps are used to move samples and the tissues frequently develop a 'charge', causing them to stick to the forceps. A cooled wooden cocktail stick can be used to dislodge them, or ceramic forceps can be used if necessary. The vials of frozen tissue and substitution material are transferred to a substitution vessel where the temperature can be controlled and a nitrogen atmosphere can be maintained.

Once the samples are in the substitution vessel, the temperature is raised typically at 5 °C/h to −90 °C, and this temperature is maintained for 24 h. This temperature is cold enough to prevent recrystallization of water and thus tissue disruption, but high enough for movement of water to take place and allow substitution with the liquid methanol. After 24 h, approximately 90% of the water has been substituted. The substitution medium is replaced and the temperature is raised to −70 °C for 24 h. The substitution medium is changed again and the temperature is raised to −50 °C. The tissue is impregnated with Lowicryl HM20 over a period of 1–5 days and the resin is polymerized by UV irradiation at −50 °C. This method has been used successfully to localize adhesion molecules [34], which are notoriously labile during fixation and embedding. This is the simplest and most versatile of the post-embedding procedures.

Ultrathin-Thawed Cryosections

In this technique, the chemically fixed sample is sectioned at low temperature, thaw-mounted on to film grids, immunochemically stained, contrast counterstained and embedded/encapsulated *in situ* on the grid. Applying immunogold reagents to sections of lightly fixed tissue, free of embedding medium, can be a very sensitive method

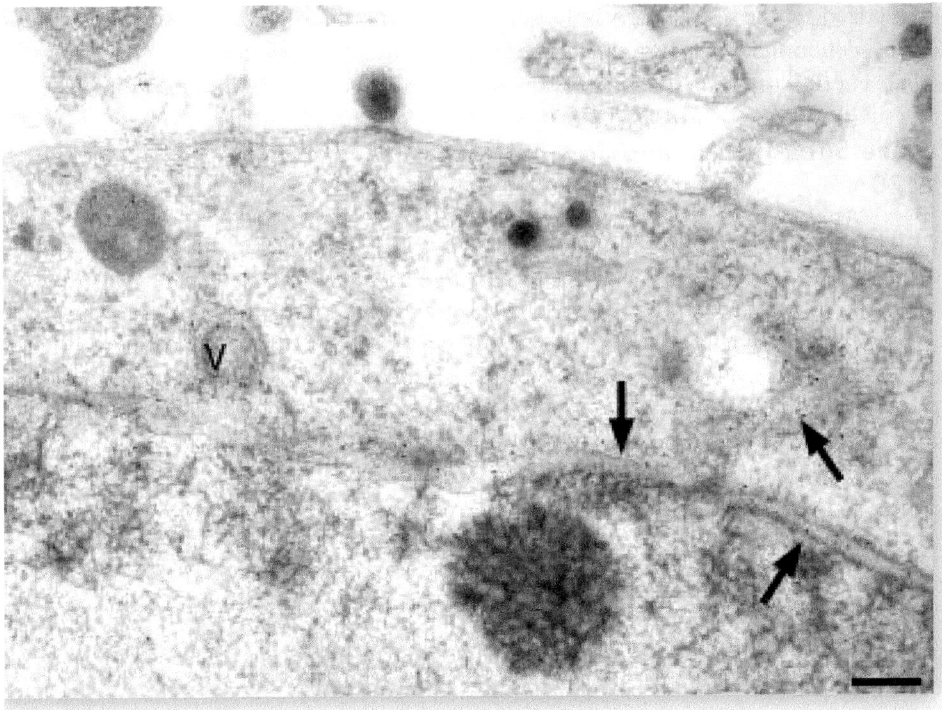

FIGURE 8.5 Thin Section of a Vero Cell Infected with Human Papilloma Virus. Cells Were Quench Frozen in Melting Propane Cooled in Liquid Nitrogen, Dehydrated by Freeze Substitution against Pure Methanol Containing 0.1% Uranyl Acetate and Low-Temperature Embedded in Lowicryl HM20. Cells Were Immunostained for Glycoprotein D. Gold Particles Indicate the Nuclear Membrane and the Rough Endoplasmic Reticulum (Arrows) and the Membrane Acquired by a Virus Particle (V) That Has Just Budded through the Nuclear Envelope. Bar, 200nm

of immunochemical staining. This technique is frequently referred to as the Tokuyasu technique [15, 37–39] after its pioneer. It is one of the few methods that is consistently used for immunochemical staining of sparse and labile membrane-bound proteins such as receptor molecules. It has been used to a great advantage in the study of receptor internalization and endocytosis [40, 41]. However, to get really good preservation of membranes, it is necessary to use 6–8% formaldehyde and many antigens may lose their ability to bind their respective antibodies after such strong fixation.

Small cubes of fixed tissue (<0.25 mm^3) impregnated with either 2.3 M sucrose [15] or a mixture of 1.9–2.1 M sucrose and 10% (w/v) polyvinylpyrrolidone [39] (to act as cryoprotectants) are mounted on pins and frozen in liquid nitrogen. The high concentrations of cryoprotectant make rapid freezing unnecessary. The frozen samples are transferred to an ultramicrotome with a cryochamber and sectioned at a temperature between −90 and −130 °C. In tissues with voids such as blood vessels, the sucrose tends to crumble rather than section. This effect is particularly noticeable in fragile embryos. It can be prevented by filling the lumen of blood vessels with gelatin [42] or by infiltration of embryos with polyacrylamide gel [38].

Thin sections are manoeuvred into position away from the cutting edge of the knife with an eyelash and retrieved on a drop of cold 2.3 M sucrose, or sucrose and polyvinylpyrrolidone, held in a copper loop with an internal diameter of 1–1.5 mm. The droplet of sucrose is moved rapidly towards the sections, which will jump towards it and 'disappear'. The sucrose must remain liquid while the sections are retrieved or they will not fully decompress, so speed is fundamental. Recently introduced alternative retrieval fluids are a 50:50 mixture of 2% (w/v) methylcellulose and 2.3 M sucrose or 1.5–2% (w/v) methyl cellulose, and 0.3–3% (w/v) uranyl acetate. If it is possible to use strong fixation (4–8%, w/v, formaldehyde for 2–4 h), the preservation of cell membranes is excellent [16] (see Fig. 8.6). These sections can be stored on buffer at 4 °C for several hours or even

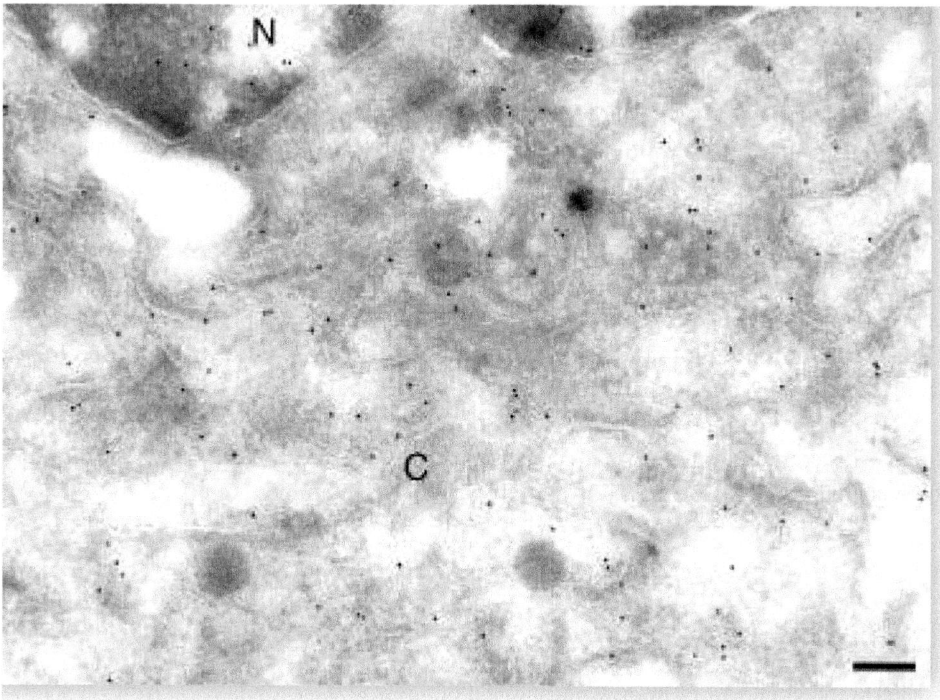

FIGURE 8.6 Ultrathin-Thawed Cryosection of Placental Syncytium. Sections Were Fixed in 6% Formaldehyde, Cryoprotected in Sucrose and Polyvinylpyrrolidone and Retrieved from the Microtome on Methylcellulose and Sucrose. Cells Were Immunostained for Copper/Zinc Superoxide Dismutase. Gold Particles Can Be Seen Over Both the Cytoplasm (C) and the Nucleoplasm (N). Bar, 200 nm

overnight, if more blocks are to be sectioned. However, caution should be exercised if the tissue has been fixed very lightly because the ultrastructure will deteriorate as a function of the time that the section is floated on the buffer.

Immunogold staining is carried out essentially the same as for resin sections with a few modifications. As the tissue has been fixed and sectioned directly after cryoprotection, residual reactive aldehyde groups may remain in the sections. These are quenched by exposure to 0.1–1% (w/v) lysine or glycine in PBS or TBS for 10 min. The absence of an embedding medium means that many primary antibodies will bind strongly after 0.5–2 h exposure, so it is often convenient to use more dilute primary antibody solutions and stain immunochemically overnight. This conserves antibodies and tends to produce less non-specific background staining.

After immunochemical staining, the sections are contrast counterstained and encapsulated in a matrix to prevent gross collapse of the section caused by surface tension effects during subsequent air drying.

The most commonly used stain is uranyl acetate, which produces a negative contrast. Many variations and alternatives to uranyl acetate have been proposed. These have been discussed in detail by Griffiths [6]. The simplest method is to rinse sections briefly in cold deionized water and incubate them on drops of 2% (w/v) methylcellulose and 3% (w/v) aqueous uranyl acetate in ratios varying from 9:1 to 5:1. Excess stain is blotted away with hardened filter paper and they are air-dried before viewing. Sections prepared in this way are stable for a considerable time. The thickness of the section and that of the final embedding layer influence the contrast in the transmission electron microscope. The thinnest, flattest, sections produced with diamond knives produce the best contrast between the section and antibodies conjugated to small colloidal gold particles (5–20 nm) [12]. As a further extension of this technique, Ripper et al. [43] have used cryoimmobilization followed by freeze substitution against acetone containing low concentrations of uranyl acetate, osmium and glutaraldehyde. After substitution, the tissues were rinsed free of fixative with several changes of dry acetone at −35 °C. They were then rehydrated, impregnated with sucrose and PVP, then ultrathin-thawed sections were prepared and labelled as earlier. This combination gives better ultrastructural preservation than that produced by chemical fixation with low concentrations of aldehydes at ambient temperature.

New Developments

One of the most exciting new developments in electron microscopy is the technique of serial block face imaging in the SEM. A resin embedded sample that has been sectioned for TEM is imaged in the SEM using a high-resolution backscattered electron detector (BSE) at 3 kV or below, typically with a probe current of less than 50 pA. This ensures the interaction volume that the BSE electrons originate from is less that the thickness of a thin section (50 nm or less). An image is generated that is identical to that of a thin section with a resolution of slightly better than 4 nm. After the first image is acquired, a diamond knife (Gatan 3View) is used to plane a slice from the surface and a second image is acquired. This process can continue for tens or even hundreds of micrometres generating stacks of images of serial sections with near-perfect registration, in a similar manner to the acquisition of *z* stacks in the confocal microscope but at TEM resolution. Serial block-face imaging can also be performed in the dual-beam focused ion beam SEM. Instead of physically planning a section from the surface of the block face, a beam of focused gallium ions is used to erode as little as a few nanometres from the block face after imaging. These techniques are reviewed in Hughes et al. [44] and Knott et al. [45].

It will be more of a challenge to combine this technique with immunochemical labelling. The labelling will have to be some variant of pre-embedding label. Therefore, chemically fixed tissues need to be permeabilized with low concentrations of detergents or by dehydrating to 100% ethanol and repeatedly freezing and thawing in liquid nitrogen. The former extracts soluble components of the cytoplasm and the latter extracts some lipids and induces 'microfractures' in the tissues allowing large molecules to penetrate. The rate-limiting factor is the size of the colloidal gold particle attached to the secondary antibody. Even 5-nm-diameter gold particle will penetrate only a few micrometres into cells or tissue. The most successful method has been to use very tiny colloidal gold particles, and silver enhance them to make them large enough to be seen easily in the TEM section or in this case at the block face. Silver enhancement may not be the most appropriate choice as prolonged treatment with osmium ferricyanide and osmium tetroxide is required to generate sufficient contrast for BSE imaging at a low voltage. This is unfortunate as osmium dissolves and extracts osmium. An alternative may be to use gold enhancement: http://www.nanoprobes.com/applications/Pre-Embedding-Labeling-Protocol-with-Gold Enhancement.html.

Gold is very stable in the prolonged staining that is required for block-face imaging. Cells or tissue slices are successively treated with osmium ferricyanide or ferrocyanide, a ligand such as thiocarbohydrazide or tannic acid, osmium tetroxide, uranyl acetate and hot lead aspartate. The combination of particulate immunostaining with automated block-face imaging would be a very powerful technique.

Another development that may lend itself to block-face imaging is the photo-oxidation of diaminobenzidine (DAB) by miniSOG, a technique that has emerged in the last few years [46]. The insoluble polymers of DAB are highly osmiophilic and should be readily visible in block-face imaging.

Recommended Protocols

Protocol 1 – Immunogold staining of epoxy resin sections

Equipment and reagents

- 400 mesh nickel grids
- 1% (w/v) coldwater-fish gelatin[a] in PBS or TBS containing 0.001% (v/v) Tween-20 and Triton X-100[b] (PBSG/TBSG)
- Dental wax[c] (or parafilm)

- Diamond trim tool and 45° ultra-diamond knife (Diatome AG)
- EM UCT ultramicrotome (Leica Microsystems)
- FEI Tecnai 120 TEM
- Lead citrate
- 50% (v/v) methanol
- 50% (v/v) methanol containing saturated uranyl acetate
- Dulbecco's 'A' PBS (pH 7.6)
- 1% (w/v) aqueous periodic acid[d]
- Primary antibodies optimally diluted in PBSG or TBSG
- Secondary antibodies optimally diluted in PBSG or TBSG and raised against the species of the primary antibody and conjugated to 10 or 15-nm colloidal gold particles
- 4% (w/v) aqueous sodium metaperiodate[e]
- Ultrapure water

Method

Carry out all nickel grid incubations/rinses on dental wax or parafilm strips

1. Cut thin sections of 50–70 nm and mount on to nickel grids.
2. Incubate sections on drops of 4% (w/v) aqueous sodium metaperiodate for 10 min at room temperature.
3. Rinse grids in ultrapure water for 30–40 s.
4. Incubate sections on drops of 1% (w/v) aqueous periodic acid for 10 min.
5. Rinse in ultrapure water for 30–40 s.
6. Incubate sections on drops of PBSG or TBSG for 10 min.
7. Incubate sections overnight on drops of pbsg or tbsg containing optimally diluted primary antibodies.
8. Rinse sections on 10× 100 μL drops of pbs or tbs for 2 min on each drop.
9. Incubate sections on drops of PBSG or TBSG containing optimally diluted species-specific secondary antibodies conjugated to 10 or 15-nm gold particles at room temperature for 2 h.
10. Rinse sections in ultrapure water for 30–40 s.
11. Counterstain sections by floating grids section side down on drops of 50% (v/v) methanol containing saturated uranyl acetate for 0.5–10 min at room temperature, followed by a rinse in 50% (v/v) methanol and a rinse in ultrapure water [47].
12. Counterstain sections by floating grids section side down on drops of lead citrate [48] for 0.5–10 min at room temperature in a petri dish containing a few grains of moistened potassium hydroxide (to prevent lead carbonate precipitation).
13. Rinse grids extensively in ultrapure water and view at 80–120 kv in a transmission electron microscope.

Notes

[a]Used as a competitive inhibitor of non-specific staining.
[b]Used as a detergent to facilitate access of antibody to antigen.
[c]Used as a clean hydrophobic surface to perform immunogold staining of thin sections mounted on TEM grids and floated on small drops of reagents.
[d]Used as an oxidizing agent to remove osmium tetroxide from the surface of thin sections. In some cases, this will enhance the binding of an antibody to its antigen at that surface.
[e]Used to remove osmium tetroxide from the surface of the thin section.

Protocol 2 – Immunogold staining of LR White resin sections

Please refer to the relevant notes in Protocol 1.

Equipment and reagents

- 400 mesh nickel film grids
- Aluminium weighing boats or Kapsto GPN600 polyethylene or polypropylene caps
- 1% (w/v) coldwater-fish gelatin[a] in PBS or TBS containing 0.001% (v/v) Tween-20 and Triton X-100[b] (PBSG/TBSG)
- Dental wax[c] (or parafilm)
- Diamond trim tool and 45° ultra-diamond knife (Diatome AG)
- EM UCT ultramicrotome (Leica Microsystems)
- 70% (v/v) ethanol
- 95% (v/v) ethanol
- 100% ethanol
- FEI Tecnai 120 transmission electron microscope
- 4% (w/v) formaldehyde (made from freshly depolymerized paraformaldehyde) in 0.1 M PIPES buffer (pH 7.4) containing 2 mM/L calcium chloride[f]
- Gelatin capsules
- Lead citrate
- 50:50 mixture of 100% LR White resin (hard consistency) and 100% ethanol
- Melinex polyester sheet
- 50% (v/v) methanol
- 50% (v/v) methanol containing saturated uranyl acetate
- Dulbecco's 'A' PBS (pH 7.6)
- 0.1 M PIPES buffer (pH 7.4)
- Primary antibodies optimally diluted in PBSG or TBSG
- Secondary antibodies optimally diluted in PBSG or TBSG and raised against the species of the primary antibody and conjugated to 10 or 15-nm colloidal gold particles
- 0.9% (w/v) sodium chloride
- Ultrapure water
- 50% (v/v) methanol saturated with uranyl acetate

Method

1. Rinse cells or small pieces of tissue twice in 0.9% (w/v) sodium chloride.
2. Incubate in 4% (w/v) formaldehyde (made from freshly depolymerized paraformaldehyde) in 0.1 M PIPES buffer (pH 7.4) containing 2 mM/L calcium chloride for 1 h at 4 °C. If the cells are adherent, scrape them free from the substrate and transfer to 1.5 mL tubes.
3. Rinse cells or small pieces of tissue four times in 0.1 M PIPES buffer over a period of 20 min and twice in ultrapure water.
4. Incubate cells or small pieces of tissue in 2% (w/v) aqueous uranyl acetate for 30 min at room temperature and rinse three times in ultrapure water.
5. Dehydrate cells or small pieces of tissue in three changes of 70% (v/v) ethanol, three changes of 95% (v/v) ethanol and three changes of 100% ethanol, all for 5 min each.
6. Incubate cells or small pieces of tissue in a 50:50 mixture of 100% LR White and 100% ethanol overnight at room temperature and in two daily changes of 100% LR White.
7. Deoxygenate fresh resin under vacuum or by bubbling dry nitrogen gas through it for 1–2 min.
8. Place the cells or tissue in a gelatin capsule or an aluminium weighing boat.
9. Add enough resin to generate a positive meniscus and cover with a piece of Melinex sheet to exclude oxygen.
10. Incubate at 55 °C for 24 h to polymerize the resin.
11. Cut thin sections of 50–70 nm and mount on to nickel grids. Carry out all nickel grid incubations/rinses on dental wax or parafilm strips
12. Incubate sections on drops of PBSG or TBSG for 10 min.
13. Incubate sections on drops of PBSG or TBSG containing optimally diluted primary antibodies overnight.
14. Rinse sections on 10× 100 μL drops of PBS or TBS for 2 min on each drop.
15. Incubate sections on drops of PBSG or TBSG containing optimally diluted species-specific secondary antibodies conjugated to 10 or 15 nm gold particles at room temperature for 2 h.
16. Rinse sections in ultrapure water for 30–40 s.
17. Counterstain sections by floating grids section side down on drops of 50% (v/v) methanol saturated with uranyl acetate for 0.5–10 min at room temperature, followed by a rinse in 50% (v/v) methanol and a rinse in ultrapure water [47].
18. Counterstain sections by floating grids section side down on drops of lead citrate [48] for 0.5–10 min at room temperature in a Petri dish containing a few grains of moistened potassium hydroxide (to prevent lead carbonate precipitation).
19. Rinse grids extensively in ultrapure water and view at 80–120 kV in a transmission electron microscope.

Additional note

[f]Used to enhance the retention of phospholipids during primary fixation.

Protocol 3 – Immunogold staining following freeze substitution and low temperature embedding, after chemical fixation or after cryoimmobilization

Please refer to the relevant notes in Protocols 1 and 2.

Equipment and reagents

- 400 mesh nickel film grids
- Automated freeze substitution system (Leica Microsystems)
- 1% (w/v) coldwater-fish gelatin[a] in PBS or TBS containing 0.001% Tween-20 and Triton X-100[b] (PBSG/TBSG)
- Dental wax[c] (or parafilm)
- Diamond trim tool and 45° ultra-diamond knife (Diatome AG)
- EM UCT ultramicrotome, automated freeze substitution device and CPC (cryo-prep centre) freezing station (Leica Microsystems)
- FEI Tecnai 120 transmission electron microscope
- 4% (w/v) formaldehyde (made from freshly depolymerized paraformaldehyde) in 0.1 M PIPES buffer (pH 7.4) containing 2 mMol/L calcium chloride[f]
- Lead citrate
- 50:50 mixture of 100% methanol and 100% HM20 resin[g]
- 50% (v/v) methanol
- 50% (v/v) methanol containing saturated uranyl acetate
- 100% methanol containing 0.05% (w/v) uranyl acetate
- Dulbecco's 'A' PBS (pH 7.6)
- 0.1 M PIPES buffer (pH 7.4)
- Primary antibodies optimally diluted in PBSG or TBSG
- Secondary antibodies optimally diluted in PBSG or TBSG and raised against the species of the primary antibody and conjugated to 10 or 15 nm colloidal gold particles
- 30% (v/v) polypropylene glycol in PBS or TBS (add 1% (w/v) BSA if cells are the subject)
- 0.9% (w/v) sodium chloride
- Ultrapure water

Method

1. Rinse cells or small pieces of tissue twice in 0.9% sodium chloride.
2. Incubate cells or small pieces of tissue in 4% (w/v) formaldehyde (made from freshly depolymerized paraformaldehyde) in 0.1 M PIPES buffer (pH 7.4) containing 2 mMol/L calcium chloride[f] for 1 h at 4 °C. If the cells are adherent, scrape them free from the substrate and transfer to 1.5 mL tubes.
3. Rinse cells or small pieces of tissue four times in 0.1 M PIPES buffer over a period of 20 min and twice in ultrapure water.
4. Incubate cells or small pieces of tissue in 30% (v/v) polypropylene glycol in PBS (add 1% (w/v) BSA if cells are the subject) at room temperature for 2 h.

5. If the subjects are cells, spin down to concentrate them, aspirate off the medium and transfer them to a small piece of aluminium foil. If the subject is a small piece of tissue, drain it and transfer it to foil.
6. Freeze the cells or small pieces of tissue in liquid propane cooled in liquid nitrogen. If the subject is to be cryoimmobilized without chemical fixation, freeze it by impact against a gold-coated copper block.
7. Transfer cells or small pieces of tissue to the automated freeze substitution system and incubate for 24 h at −90 °C in pure methanol containing 0.05% (w/v) uranyl acetate.
8. Warm cells or small pieces of tissue to −70 °C and maintain for 24 h.
9. Warm cells or small pieces of tissue to −50 °C and rinse in four changes of pure methanol over a period of 2 h.
10. Mix and deoxygenate 100% HM20 resin by bubbling dry nitrogen gas through it for 5 min.
11. Incubate cells or small pieces of tissue in a 50:50 mixture of 100% methanol and 100% HM20 resin at room temperature overnight.
12. Incubate cells or small pieces of tissue in 100% HM20 resin at room temperature for 4 days, changing the resin daily.
13. Polymerize the resin by UV irradiation for 24 h at −50 °C, 24 h at −40 °C and 48 h at 15 °C.
14. Cut thin sections of 50–70 nm and mount on to nickel grids. Carry out all nickel grid incubations/rinses on dental wax or parafilm strips.
15. Incubate sections on drops of PBSG or TBSG at room temperature for 10 min.
16. Incubate sections on drops of PBSG or TBSG containing optimally diluted primary antibodies at room temperature overnight.
17. Rinse sections on 10× 100 μL drops of PBS or TBS for 2 min on each drop.
18. Incubate sections on drops of PBSG or TBSG containing optimally diluted species-specific secondary antibodies conjugated to 10 or 15 nm gold particles at room temperature for 2 h.
19. Rinse sections in ultrapure water for 30–40 s.
20. Counterstain sections by floating grids section side down on drops of 50% (v/v) methanol containing saturated uranyl acetate for 0.5–10 min at room temperature, followed by a rinse in 50% (v/v) methanol and a rinse in ultrapure water [47].
21. Counterstain sections by floating grids section side down on drops of lead citrate [48] for 0.5–10 min at room temperature in a Petri dish containing a few grains of moistened potassium hydroxide (to prevent lead carbonate precipitation).
22. Rinse grids extensively in ultrapure water and view at 80–120 kV in a transmission electron microscope.

Additional note

[g]HM20 is a low-temperature resin, providing low viscosity at low temperatures.

Protocol 4 – Immunogold staining of ultrathin-thawed cryosections
Please refer to the relevant notes in Protocols 1 and 2.

Equipment and Reagents

- 400 mesh nickel film grids
- 1% (w/v) coldwater-fish gelatin[a] in PBS or TBS containing 0.001% Tween-20 and Triton X100[b] (PBSG/TBSG)
- Diamond trim tool and 45° ultra-cryo-diamond knife (Diatome AG)
- EM UCT ultramicrotome, frozen cryosection module (FCS) and CPC (cryo-prep centre) freezing station (Leica Microsystems)
- FEI Tecnai 120 transmission electron microscope
- 2% (w/v) formaldehyde (made from freshly depolymerized paraformaldehyde)
- 8% (w/v) formaldehyde (made from freshly depolymerized paraformaldehyde) in 0.1 M PIPES buffer (pH 7.4) containing 2 mMol/L calcium chloride[f]
- 10% (w/v) gelatin in PBS
- Hardened filter paper
- 50% (v/v) methanol
- 50% (v/v) methanol containing saturated uranyl acetate
- 2% (w/v) methylcellulose[h] and 3% (w/v) aqueous uranyl acetate in ratios varying from 9:1 to 5:1
- Dulbecco's 'A' PBS (pH 7.6)
- 0.1 M PIPES buffer (pH 7.4)
- Primary antibodies optimally diluted in PBSG or TBSG
- Secondary antibodies optimally diluted in PBSG or TBSG and raised against the species of the primary antibody and conjugated to 10 or 15-nm colloidal gold particles
- 0.9% (w/v) sodium chloride
- 50:50 mixture of 2.3 M sucrose and 2% (w/v) methylcellulose[h]
- 1.9 M sucrose[i] and 10% (w/v) polyvinylpyrrolidone (PVP-10)[i]
- Ultrapure water

Method

1. Rinse cells or small pieces of tissue twice in 0.9% (w/v) sodium chloride.
2. Incubate in 8% (w/v) formaldehyde (made from freshly depolymerized paraformaldehyde) in 0.1 M PIPES buffer (pH 7.4) containing 2 mM/L calcium chloride[f] for 1 h at 4 °C. If the cells are adherent, scrape them free from the substrate and transfer to 1.5 mL tubes.
3. Rinse four times in 0.1 M PIPES buffer at room temperature over a period of 20 min and twice in ultrapure water.

4. Incubate in 10% (w/v) gelatin in PBS for 2 h and spin down to form a pellet. Cool to 4 °C and fix in 2% (w/v) formaldehyde for 2 h. Small pieces of fixed tissue can be trimmed to 0.5 mm in one dimension.
5. Trim to 0.5 mm^3 cubes and incubate in 1.9 M sucrose[i] and 10% (w/v) PVP-10[i] overnight at 4 °C.
6. Freeze the cubes on to sectioning pins in liquid nitrogen. Transfer to the frozen cryosection module and cut thin sections of 90–140 nm.
7. Place a drop of a 50:50 mixture of 2.3 M sucrose and 2% (w/v) methylcellulose in a 1-mm-diameter loop and retrieve the sections from the frozen cryosection module. Allow the sucrose to thaw at room temperature and touch the sections on to the Formvar surface of a film grid.
8. Transfer to drops of PBSG or TBSG.
9. Incubate sections on drops of PBSG or TBSG containing optimally diluted primary antibodies at room temperature overnight.
10. Rinse sections on 10× 100 μL drops of PBS or TBS for 2 min on each drop.
11. Incubate sections on drops of PBSG or TBSG containing optimally diluted species-specific secondary antibodies conjugated to 10 or 15 nm gold particles at room temperature for 2 h.
12. Rinse sections on 10× 100 μL drops of PBS or TBS for 2 min on each drop.
13. Rinse briefly in ultrapure water and incubate sections on drops of 2% (w/v) methylcellulose and 3% (w/v) aqueous uranyl acetate in ratios varying from 9:1 to 5:1, until the desired contrast is achieved. Blot away excess stain using hardened filter paper and air dry sections before viewing at 80 or 120 kV in a transmission electron microscope.

Additional notes

[h]Used as an embedding medium to prevent the sections collapsing totally during air drying.
[i]Used as a cryoprotectant.

TBS can be substituted for PBS in some cases less background labelling is found when TBS is used and in other cases the reverse happens.

In all these examples, PIPES is used as the buffer of choice for fixatives. HEPES or sodium cacodylate buffers can be used as alternatives if preferred.

REFERENCES

1. Coons, A.H., Creech, H.J. and Jones, R.N. (1941) The first practical demonstration of tagging an antibody with a fluorochrome. *Proceedings of the Society for Experimental Biology and Medicine*, **47**, 200–202.
2. Singer, S.J. (1959) The first practical demonstration of tagging an antibody with ferratin. *Nature*, **183**, 1523–1524.
3. Rifkind, R.A., Hsu, K.C. and Morgan, C. (1964) Immunochemical staining for electron microscopy. *Journal of Histochemistry & Cytochemistry*, **12**, 131–136.
4. Faulk, W.P. and Taylor, G.M. (1971) An immunocolloid method for the electron microscope. *Immunochemistry*, **8**, 1081–1083.

5. Frens, G. (1973) Controlled nucleation for regulation of particle-size in monodisperse gold suspensions. *Nature Physical Science*, **241**, 20–22.
6. Griffiths, G. (1993) A comprehensive description of the ultrathin thawed cryosection technique of immunolabeling, in *Fine Structure Immunocytochemistry* (ed. G. Griffiths), Springer-Verlag, Heidelberg, Germany.
7. Lucocq, J. (1993) A succinct description of quantification of immunolabeling using unbiased stereological techniques. *Trends in Cell Biology*, **3**, 354–358.
8. Storm-Mathisen, J. and Ottersen, O.P. (1990) A description of how to correlate gold label density with antigen concentration. *Journal of Histochemistry & Cytochemistry*, **38**, 1733–1743.
9. Hayat, M.A. (1981) A thorough description of the chemistry of fixation for electron microscopy, in *Fixation for Electron Microscopy* (ed. M.A. Hayat), Academic Press, NY, USA.
10. Glauert, A.M. and Lewis, P.R. (1998) A thorough description of the chemistry of fixation for electron microscopy and choice of embedding media, in *Practical Methods in Electron Microscopy, Vol. 17: Biological Specimen Preparation for Transmission Electron Microscopy* (eds A.M. Glauert and P.R. Lewis), Portland Press, London, UK.
11. Bendayan, M., Nanci, A. and Kan, F.W.K. (1987) Effect of tissue processing on colloidal gold cytochemistry. *Journal of Histochemistry & Cytochemistry*, **35**, 983–996.
12. Peters, P.J., Mironov, A. Jr., Peretz, D. *et al.* (2003) A superb example of the use of the ultrathin thawed cryosection technique of immunolabeling. *The Journal of Cell Biology*, **162**, 703–717.
13. Skepper, J.N., Woodward, J.M. and Navaratnam, V. (1988) Immunocytochemical localization of natriuretic peptide sequences in the human right auricle. *Journal of Molecular and Cellular Cardiology*, **20**, 343–353.
14. Crapo, J.D., Oury, T., Rabouille, C. *et al.* (1992) An example of the apparently paradoxical requirement to use strong fixation to retain and label soluble antigens. *Proceedings of the National Academy of Sciences of the United States of America*, **89**, 10405–10409.
15. Tokuyasu, K.T. (1986) Application of cryoultramicrotomy to immunocytochemistry. *Journal of Microscopy*, **143**, 139–149.
16. Liou, W., Geuze, H.J. and Slot, J.W. (1996) A review of modifications to the ultrathin thawed cryosection techniques that significantly improve ultrastructural preservation. *Histochemistry and Cell Biology*, **106**, 41–58.
17. Chung, D., Kim, E. and So, P.T. (2006) Extended resolution wide-field optical imaging: objective-launched standing-wave total internal reflection fluorescence microscopy. *Optics Letters*, **31**, 945–947.
18. Egner, A., Verrier, S., Goroshkov, A. *et al.* (2004) 4Pi- microscopy of the Golgi apparatus in live mammalian cells. *Journal of Structural Biology*, **147**, 70–76.
19. Willig, K.I., Rizzoli, S.O., Westphal, V. *et al.* (2006) STED microscopy reveals that synaptotagmin remains clustered after synaptic vesicle exocytosis. *Nature*, **440**, 935–939.
20. Brændgaard, H. and Gundersen, H.J.G. (1986) An example of the 'reference trap' that can invalidate quantification. *Journal of Neuroscience Methods*, **18**, 39–78.
21. Howard, C.V. and Reed, M.G. (1998) A detailed description, with practical examples, of how to apply stereology, in *Unbiased Stereology* (eds C.V. Howard and M.G. Reed), BIOS Scientific Publishers, Oxford, UK.
22. Gundersen, H.J.G. (1977) The forbidden line rule. *Journal of Microscopy*, **111**, 219–223.
23. Lucocq, J.M., Habermann, A., Watt, S. *et al.* (2004) A rapid method for assessing the distribution of gold labeling on thin sections. *Journal of Histochemistry & Cytochemistry*, **52**, 991–1000.
24. Mayhew, T.M., Lucocq, J.M. and Griffiths, G. (2002) Nonparametric methods of quantifying immunogold labeling. *Journal of Microscopy*, **205**, 153–164.
25. Skepper, J.N. (2000) A comprehensive review of the methods available for immunolabeling and their strengths and weaknesses. *Journal of Microscopy*, **199**, 1–36.

26. Causton, B. (1984) The choice of resins for electron immunocytochemistry, in *Immunolabelling for Electron Microscopy* (eds J.M. Polak and I.M. Varndell), Elsevier, Amsterdam, pp. 29–36.
27. Bendayan, M. and Zollinger, M. (1983) Ultrastructural localization of antigenic sites on osmium-fixed tissues applying the protein A-gold techniques. *Journal of Histochemistry & Cytochemistry*, **31**, 101–109.
28. Probert, L., de May, J. and Polak, J. (1981) Distinct subpopulations of enteric p-type neurones contain substance P and vasoactive intestinal polypeptide. *Nature*, **294**, 470–471.
29. Bendayan, M., Nanci, A., Herbener, G.H. *et al.* (1986) A review of the study of protein secretion applying the protein A-gold immunocytochemical approach. *American Journal of Anatomy*, **175**, 379–400.
30. Varndell, I.M., Sikri, K.L., Hennessy, R.J. *et al.* (1986) Somatostatin-containing D cells exhibit immunoreactivity for rat somatostatin cryptic peptide in six mammalian species - An electron-microscopical study. *Cell and Tissue Research*, **246**, 197–204.
31. Newman, T.M., Severs, N.J. and Skepper, J.N. (1991) The pathway of atrial natriuretic peptide release – from cell to plasma. *Cardioscience*, **2**, 263–272.
32. Causton, B. (1981) The introduction of LR White. *Proceedings of the Royal Microscopical Society*, **16**, 265–271.
33. Berryman, M.A. and Rodewald, R.D. (1990) A method for improving membrane preservation during embedding in acrylic resin. *Journal of Histochemistry & Cytochemistry*, **38**, 159–170.
34. Zajicek, J., Wing, M., Skepper, J. and Compston, A. (1995) Human oligodendrocytes are not sensitive to complement: a study of CD59 expression in the human central nervous system. *Laboratory Investigation*, **73**, 128–138.
35. Monaghan, P. and Robertson, D. (1990) A method for cryoimmobilization, freeze substitution, and low-temperature embedding without chemical fixation. *Journal of Microscopy*, **158**, 355–363.
36. Skepper, J.N., Whiteley, A., Browne, H. and Minson, A. (2001) Herpes simplex virus nucleocapsids mature to progeny virions by an envelopment → deenvelopment → reenvelopment pathway. *Journal of Virology*, **75**, 5697–5702.
37. Tokuyasu, K.T. (1973) An early review on the ultrathin thawed cryosection techniques by its pioneer. *The Journal of Cell Biology*, **57**, 551–565.
38. Tokuyasu, K.T. (1983) Present state of immunocryoultramicrotomy. *Journal of Histochemistry & Cytochemistry*, **31** (Suppl. 1A), 161–167.
39. Tokuyasu, K.T. (1989) Application of cryoultramicrotomy to immunocytochemistry. *Histochemical Journal*, **21**, 163–171.
40. Liou, W., Geuze, H.J., Geelen, M.J.H. and Slot, J.W. (1997) The autophagic and endocytic pathways converge at the nascent autophagic vacuoles. *The Journal of Cell Biology*, **136**, 61–70.
41. Klumperman, J., Kuliawat, R., Griffith, J.M. *et al.* (1998) Mannose 6-phosphate receptors are sorted from immature secretory granules via adaptor protein AP-1, clathrin, and syntaxin 6-positive vesicles. *The Journal of Cell Biology*, **141**, 359–371.
42. Russell, F.D., Skepper, J.N. and Davenport, A.P. (1998) Human endothelial cell storage granules: a novel intracellular site for isoforms of the endothelin converting enzyme. *Circulation Research*, **83**, 314–321.
43. Ripper, D., Schwarzt, H. and Stierhof, Y. (2008) Cryo-section immunolabeling of difficult to preserve specimens: Advantages of cryofixation, freeze-substitution and rehydration. *Biology of the Cell*, **100**, 109–123.
44. Hughes, L., Hawes, C., Monteith, S. and Vaughan, S. (2014) Serial block face scanning electron microscopy--the future of cell ultrastructure imaging. *Protoplasma*, **251**, 395–401.

45. Knott, G., Marchman, H., Wall, D. and Lich, B. (2008) Serial section scanning electron microscopy of adult brain tissue using focused ion beam milling. *The Journal of Neuroscience*, **28** (12), 2959–2964.
46. Butko, M.T., Yang, J., Geng, Y. *et al.* (2012) Fluorescent and photo-oxidizing TimeSTAMP tags track protein fates in light and electron microscopy. *Nature Neuroscience*, **15**, 1742–1753.
47. Gibbons, I.R. and Grimstone, A.V. (1960) On flagellar structure in certain flagellates. *The Journal of Biophysical and Biochemical Cytology*, **7**, 697–716.
48. Reynolds, E.S. (1963) The use of lead citrate at high ph as an electron-opaque stain in electron microscopy. *The Journal of Cell Biology*, **17**, 208–212.

Index

A

Abbe's law, 176
acetone fixative, 42
adhesive mounting media, 76
affinity purification chromatogram, 17
aldehyde fixatives, 44
aldehyde-fixed tissue, 210
aldehyde-induced auto-fluorescence, 146–147
Alexa Fluor® dyes, 30–32, 178
alkaline phosphatase (AP)
  chromogens, 28
  endogenous peroxidases and phosphatases, 29
  naphthol phosphate groups, hydrolyse, 28
amino acid properties, 9
antibodies, 200
  adaptive immunity, 1
  antibody production
    monoclonal, 14–15
    polyclonal, 13–14
    subcutaneous and intradermal, 12
  epitope prediction tools, 7
  immunogens types, 6–7
  labelling
    enzyme reporters, 23
    fluorescent reporters, 22–23
  peptide immunogen design, 7–11
  protein structure, 6
  purification
    affinity purification, 17–18
    ammonium sulphate precipitation, 16
    protein L, 19–20
    proteins A and G resins, 18–19
  stability and storage, 23–24
  structure, 2–6
antigen (antibody generator), 1
antigen–antibody binding, 200–204
  see also fixation
antigen concentration, 200
antigenicity indices, 7
antigen retrieval, 111–112
  conditions of, 50–51
  description, 50
  enzymatic, 54–56
  heat-induced, 51–54
  heat-induced epitope, 131–134
  proteolytic (enzymatic), 130–131
aqueous mounting media, 75–76
array stainers, 160, 161
auto-fluorescence, 144–147
automated block-face imaging, 215
automated immunochemical staining
  advantages, 147
  automated reagent handling, 148
  disadvantages, 147
  gradient staining, 148
  inadequate reagent flow, 149
  reagent precipitate, 148

Immunohistochemistry and Immunocytochemistry: Essential Methods, Second Edition. Edited by Simon Renshaw.

automated immunochemical staining (*Continued*)
stainers, 160–161
tissue damage, 148
troubleshooting, 149–154
automated immunochemistry, 157 *see also* immunochemistry
analysis algorithms, 167
antigen retrieval, 162
automated antigen retrieval systems, 164–165
automated immunochemical stainers, 160–161
bar-coding, 163
collaborative projects, 167
dewaxing, 164–165
digital pathology, 165–167
failure safeguards, 163
heating methods, 162
integrate and manage, 166
laboratory platform, choice of, 158–159
reagent application method, 161–162
scanner software, 165
slide heating, 162
slide-staining systems, 159
system running costs, 163
user friendly platforms, 162
view images, 166–167
Avian immunoglobulins, 20
avidin–biotin complex (ABC) immunochemical staining protocol, 64–68

B

Basic Local Alignment Search Tool (BLAST) algorithm, 9, 10
beta-galactosidase (β-GAL), 29
B5 fixative, 43
biotin, 143–144
Bouin's fixative, 43
bovine serum albumin (BSA), 10, 11
5-bromo-4-chloro-3-indolyl-*β*-D-galacto-pyranoside (BCIG or X-gal), 29

C

catonized BSA (cBSA), 10
Cavalieri method, 206
cell surface receptors (Fc), 4
class and subclass-specific secondary antibodies, 112–113
closed systems, staining platforms, 161
collagen auto-fluorescence, 145
colloidal gold technology, 199
compact polymer, 63
compact polymer immunochemical staining protocol, 69–71
complementarity-determining regions (CDR), 4
confocal microscopy
advantages, 169
applications, 173
bioscience research, 169
depth penetration and resolution, 174
experimental set up
excitation and emission pathways, 176–177
image size, shape and zoom of sample, 182–183
magnification and resolution, 175
microscope set up, 175–176
multicolour confocal imaging, 177–178
multicolour experiments, 179
numerical aperture, 176
objective lens, 175
optimizing detectors, 180
PMT detector, 181
sample preparation, 174
scan average, 182
spectral bleedthrough, 178, 179
fast live imaging, 191–194
fluorescence photoactivation, 194–195

fluorescence recovery after photobleaching, 194–195
history of, 170
images
deconvolution of, 197
3D presentation, 197–198
processing for presentation, 195–196
quantification of, 197
limitations of, 174
live cell imaging, 186–191
sample collection using confocal tiling, 186
set-up, 172
three-dimensional confocal imaging, 183–186
usage of, 173
vs. widefield epifluorescent microscopy, 170–171
workflow summary, 183
working of, 171–173
confocal point-by-point scanning, 172
conjugate-activated HRP and thiolated antibody, 23
conjugating peptide immunogens to carrier protein, 11–12
conventional epi-fluorescence, 103, 105
counterstains
for enzyme/chromogen immunostaining, 71–72
for fluorescent immunostaining, 72–74
cross-linking fixatives, 125–127
Cy dyes, 178
cysteine (C), 4, 8–12, 18, 20, 21

D

DAPI, 72–74
decalcification, 49–50, 129–130
deconvolution, of confocal microscopy, 197
dewaxing, 164–165
dextran–polymer systems, 63
de-Zenkerization, 43
3,3′-diaminobenzidine (DAB) shielding, 109–110
differentiation process, 72
direct *vs.* indirect immunochemical staining method, 107
dithiothreitol (DTT), 4
double immunostaining
using conjugated Fab fragments, 113
using unconjugated Fab fragments, 114
DRAQ5™, 74
drop zone, for stainers, 162
DyLight dyes, 178
DyLight Fluor®, 30, 32

E

elastin auto-fluorescence, 145
electron microscopy, developments, 214–215
endogenous peroxidase activity, 141–142
endogenous phosphatase activity, 142–143
enzymatic antigen retrieval, 54–56
enzyme, 54, 104, 106, 107, 112
chromogen immunostaining, counterstains for, 71–72
digestion factors, 131
reporter labels, 109–110
vs. fluorescent reporter labels, 62
selection of, 61
enzyme-based retrieval method, 130
enzyme-linked immunosorbent assay (ELISA), 15
epitope retrieval *see* antigen retrieval
1-ethyl-3-[3-dimethylaminopropyl] carbodiimide hydrochloride (EDC), 11

F

Fab (fragment antigen-binding region) fragments, 19–21, 113–115

fast scanning confocal microscopes, types of, 191, 192
Fc receptors, 141
ferritin, 199
fixation, 37
  beneficial effects of, 44
  fixatives, types of, 38–44
  quality control considerations, 44–46
  safety, 202
  temperature and duration of, 202
  and tissue processing, 125–128
  of tissues and organs, 201
fixatives
  aldehyde fixatives, considerations of, 46
  classification of, 38
  concentration, 45
  cross-linking fixatives, 125–127
  duration and temperature, 45–46
  penetration, 45
  precipitant, 127
  protein-denaturing, considerations of, 46
fluorescein isothiocyanate (FITC), 22
fluorescence detection
  absorption and emission spectra, fluorochromes, 31
  fluorochrome characteristics, 30–32
  internal conversion, 29
fluorescence photoactivation, 194–195
fluorescence recovery after photobleaching (FRAP), 194–195
fluorescent immunostaining, counterstains for, 72–74
fluorescent nuclear and cell membrane counterstains, 73
Fluorescent or Forster Resonance Energy Transfer (FRET) FRAP, 193–194
fluorescent reporter labels
  description, 61
  vs. enzyme reporter labels, 62
fluorescent reporters, 22–23
fluorescent secondary antibody method, 201
fluorochromes, 30–32, 61–62
forbidden line rule, 206
formaldehyde, 200, 202 *see also* fixatives
  fixation mechanism, 39–40
  polymer formation, 38
fragment antibody preparations, 20–21
freeze substitution, in Lowicryl HM20, 211, 212

G

glucose oxidase, 29
glutaraldehyde, 40–41, 200, 202, 208
gold enhancement, 215

H

haematoxylin, 71–72
heat-induced epitope retrieval (HIER), 131
  advantages, 134
  antigen retrieval solutions, 133–134
  awareness factors, 52
  decloaker device, 133
  description, 51
  disadvantages, 134
  EDTA, 51
  hypotheses, 132
  microwave method, 53–54
  microwave oven heating method, 132
  microwave pressure cooking method, 133
  pressure cooker method, 52–53
  pressure cooking method, 132–133
  purpose-built systems, 51
  steamer method, 133
  success of, 51
  trisodium citrate buffer, 51
higher profile digital pathology solution, 167
Hoechst dyes, 74
horseradish peroxidase (HRP)
  bacteriostatic agents and chemicals, 28
  chromogens, 27
  hydrogen peroxide, 27

hydrophilicity indices, 7
hydrophobic interactions, 140–141
hypervariable (HV) loops, 4

I

2-iminothiolane (2-IT) antibody thiolation, 23
immune epitope database (IEDB), 8
immunochemical staining, 134, 199
  absorption control, 138
  antibody optimization experiment, 59–60
  avidin–biotin complex, 64–68
  avidin–biotin techniques, 135–136
  compact polymer, 69–71
  control blocks, 139
  detection systems, choice of, 61–64
  frozen sections, preparation of, 127–128
  internal controls, 139
  label-conjugated secondary antibody, 68–69
  optimal dilution of primary antibody, 136
  positive and negative tissue controls, 137
  pre-antibody purchase/optimization research, 57–59
  reagent controls, 137–138
  reporter labels, 61–62
  signal amplification, 62–64
  troubleshooting, 76
immunochemistry
  aim of, 124–125
  antigen retrieval, 130–134
  auto-fluorescence, 144–147
  biotin, 143–144
  controls, 136–139
  decalcification, 129–130
  description, 35
  endogenous peroxidase activity, 141–142
  endogenous phosphatase activity, 142–143
  Fc receptors, 141
  fixation process, 37
    beneficial effects of, 44
    fixatives, types of, 38–44
    quality control considerations, 44–46
    and tissue processing, 125–128
  hydrophobic interactions, 140–141
  immunochemical methodologies, 134–136
  ionic interactions, 141
  microscopic interpretation, 139–140
  microtomy, 128–129
  monoclonal and polyclonal antibodies for, 15–16
  negative antigen controls, 56
  positive antigen controls, 56
  reagent controls, 56–57
  specimen formats for
    cytological preparations, 37
    free-floating sections, 37
    frozen section, 36–37
    paraffin embedded, 36
immunocytochemistry/immunofluorescence
  A431 cells
    ab192055 staining cytokeratin 14 in, 80
    ab207351 staining cytokeratin 5 in, 78
  HeLa cells
    ab197240 staining alpha smooth muscle actin in, 83
    ab195887 staining alpha tubulin in, 77
    ab196158 staining calreticulin in, 81
    ab196159 staining calreticulin in, 82
    ab203410 staining fibrillarin in, 87
    ab203850 staining histone H3 in, 88
    ab205769 staining lamin A+C in, 89
    ab185036 staining MAP1LC3A in, 79
    ab201543 staining SENP1 in, 86
    ab207014 staining STAT6 in, 77
    ab199814 staining YY1 in, 85

immunocytochemistry/immunofluorescence (*Continued*)
  MCF7 cells, ab198608 staining alpha actinin 4 in, 84
immunoenzymatic detection, 25, 26
immunofluorescence (IF), 104, 105
immunoglobulins (Ig), 18
  avian, 20
  classes or isotypes, 2
  domains, 4
  immunoglobulin fold, 4
  molecule, schematic illustration of, 2, 3
immunogold staining, 214
  of epoxy resin sections, 215–216
  following freeze substitution and low temperature embedding, 219–220
  of LR White resin sections, 217–218
  of ultrathin-thawed cryosections, 221–222
immunohistochemistry
  adenosine A1 receptor staining, 92
  cardiac troponin I staining, 96
  CD7 (mature T-cell marker) staining, 91
  chromogranin A staining, 99
  cytokeratin 10 staining, 95
  lamin B1 (nuclear membrane marker) staining, 90
  LRRK2 staining, 94
  MCM7 staining, 93
  metabotropic glutamate receptor 5 staining, 98
  PIM1 staining, 100
  tissue processing, 46–47
  VCAM1 staining, 97
ion exchange, 20
ionic interactions, 141
isotype control, 138

K

Karnovsky's fixative, 41
keyhole limpet hemocyanin (KLH), 10, 11

L

label-conjugated secondary antibody immunochemical staining protocol, 68–69
label (staining) density, 205–207
Leica Aperio® high-throughput automated slide scanner, 166
Leica Biosystems Bond III, 160
light-absorbing pigment rhodopsin, 206
lipofuscin auto-fluorescence, 145–146
live cell imaging, 186
  confocal microscope with environmental incubation for, 187
  dyes, 188–189
  experiment set-up, 190–191
  fluorescent proteins, 189–190
  visualization of samples, 188–191
London Resin Gold, 208–209
London Resin White, 208, 209
low-temperature embedding, in Lowicryl HM20, 211, 212
lysine, 42

M

mammalian and avian host species, 13
memory B cells, 1, 12, 13
mercuric chloride/formaldehyde-based fixatives, 43
microtomy, 47, 128–129
monoclonal cell culture supernatants purification, 18
mordants, 72
mounting, 74
mounting medium
  characteristics, 74–75
  organic and aqueous, 75–76
multiple antigenic peptides (MAP), 11
multiple immunochemical staining techniques
  advantages and technical challenges, 103–104
  cell biology, 103

directly reporter labelled primary antibodies, 112
direct *vs.* indirect method, 107
double staining using same-species primary antibodies, 112
experiment, 106
experiment controls, 110–111
method selection, 104–106
primary and secondary antibody combination, 107–108
sequential enzymatic, 117
sequential fluorescence, 119–122
simultaneous enzymatic, 118
simultaneous fluorescence, 115–116
simultaneous *vs.* sequential staining, 106–107

N

N-and C-terminal regions, 8
NHS and iodoacetyl-activated resins, 20
N-hydroxysuccinimide (NHS), 11, 20
non-adhesive mounting media, 76
non-alcohol-containing haematoxylin, 72
non-coagulative fixatives, 44
non-specific background staining, 140–141
nuclear tinctorial counterstain, 72

O

ommatidia, 204, 205
open systems, staining platforms, 161
organic mounting media, 75–76
osmium tetroxide, 202, 203, 208–210, 215
ovalbumin (OVA), 10

P

paraffin-fixed tissues, 104
paraformaldehyde, 39, 42
peptide carrier protein
BSA, 11
cBSA, 10
MAP, 11
OVA, 10
peptide immunogen design considerations
amino acid properties, 9
candidate regions identification, 8
post-translational modification peptide immunogens, 10
potential cross-reactivity, 9–10
protein research, 7–8
periodate oxidizes sugars, 42
periodic acid, 202–203
phalloidin, 74
photoactivatable GFP (PA-GFP) proteins, 189
photoswitchable GFP (PS-GFP) proteins, 189–190
picric acid, 43
placental syncytium, ultrathin-thawed cryosection of, 213
polymer systems, 63–64
potassium dichromate, 42
precipitant fixatives, 127
primary antibodies, 15, 107, 108, 113
printer cartridge-like dispensers, 162
probe-type dispensers, 162
progressive haematoxylins, 72
proline (P), 9
protein-denaturing agents, 41–42
protein-denaturing fixatives, 46

Q

quality assessment *vs.* control, 124
quality assurance, 123
elements for, 124
quality control
definition, 123
vs. quality assessment, 124
quantification, of confocal microscopy, 197
quantum yield, 178

R

random sampling strategies, 200
reference trap phenomenon, 206
regressive haematoxylins, 72
reporter labels
  enzymatic labels, 109–110
    AP, 28–29
    β-GAL, 29
    vs. fluorescent reporter labels, 62
    HRP, 27–28
    selection of, 61
  fluorescence detection, 29–32
  fluorochrome characteristics, 30–32
  immunohistochemical, 25
rotary stainers, 161, 162

S

scanning methods, 180
secondary antibodies, 16, 107, 108
sequential scanning, 180
serial block face imaging technique, 214–215
serine (S), 8, 9
simultaneous scanning, 180
simultaneous *vs.* sequential staining, 106–107
sodium dodecyl sulphate-polyacrylamide gel electrophoresis (SDS-PAGE), 6
sodium metaperiodate, 202, 203
species-specific antibody method, 201
specimen storage
  cytological specimens, 49
  fluorescently labelled specimens, 49
  frozen tissue sections, 49
  paraffin-embedded sections
    post-immunohistochemically stained, 48–49
    pre-immunohistochemically stained, 48
spectral detectors, 177
spectral non-mixing, 167
staining intensity, 72
Standard Laser and Emission filter configurations, on confocal microscopes, 177
Staphylococcus aureus, 18
sulfo-SMCC (sulfosuccinimidyl-4-(*N*-maleimidomethyl)cyclohexane-1-carboxylate), 11–12
  HRP activation, 23
syringe-like dispensers, 162

T

tetramethylrhodamine-5-(and 6)-isothiocyanate (TRITC), 22
three-dimensional confocal image presentation, 197–198
threonine (T), 8, 9
tissue microarray (TMA), 47–48, 139
tissue processing, 124
Tokuyasu' method, 204, 212
Traut's reagent, 23
trypsin, 130
tyrosine (Y), 9

U

ultrastructural immunochemistry
  chemical fixation and embedding, in highly cross-linked epoxy resin, 208–209
  controls, 204
  electron microscopy, 204
  fixation and effect on antigen–antibody binding, 200–204
  freeze substitution, in Lowicryl HM20, 211, 212
  London Resin Gold, 208–209
  London Resin White, 208, 209
  low-temperature embedding, in Lowicryl HM20, 211, 212
  ommatidia, 204, 205

4Pi microscopy, 204
post-embedding methods, 207
pre-embedding method, 207
quantification, 205–207
stimulated emission depletion microscopy, 204
total internal reflectance microscopy, 204
ultrathin-thawed cryosections, 211–214
unmasking *see* antigen retrieval
uranyl acetate, 208, 214

W

weak and strong fixatives, effects of, 202
Western blotting, 6, 8, 58, 200
wheat germ agglutinin, 74

X

xylene-free tissue processing, 47

Z

Zenker's fixative, 43
zinc formalin fixative, 44

Printed and bound by CPI Group (UK) Ltd, Croydon, CR0 4YY
27/10/2024
14580214-0002